The Last Lions

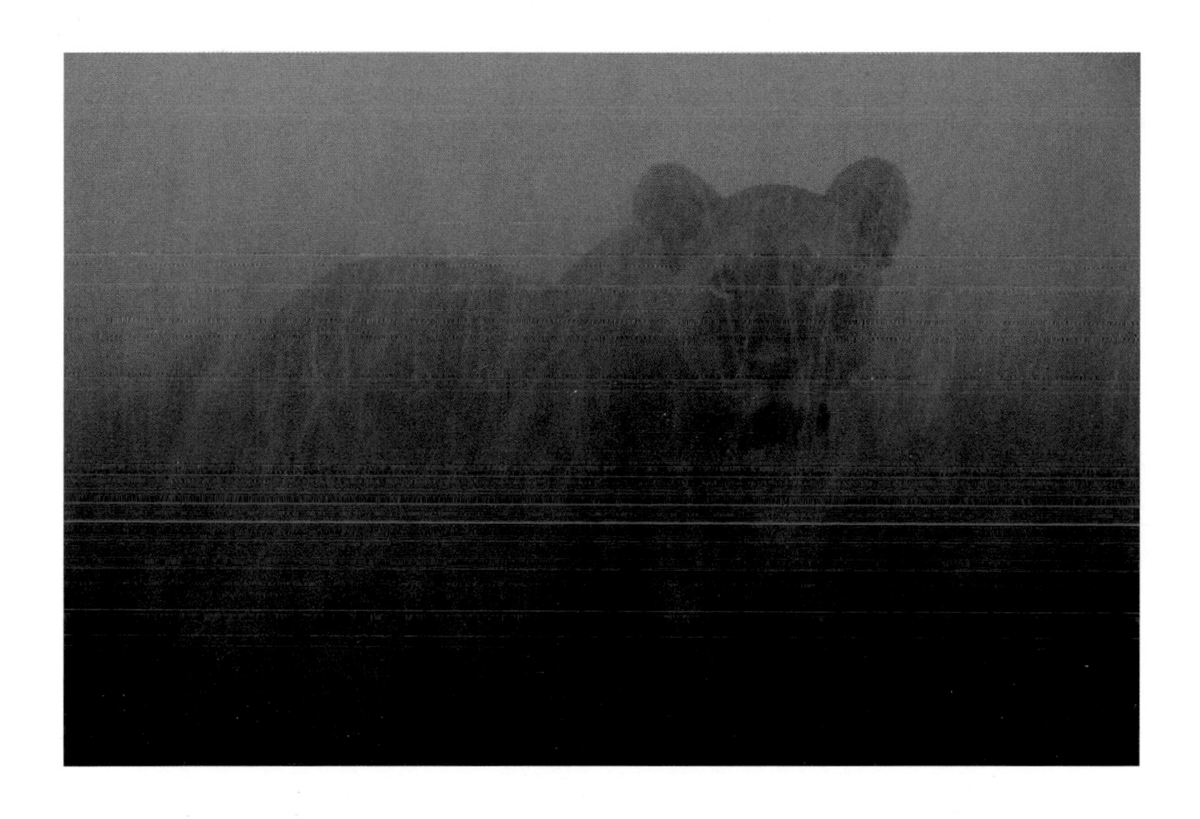

The Last Lions

PHOTOGRAPHED BY BEVERLY JOUBERT AND WRITTEN BY DERECK JOUBERT

NATIONAL GEOGRAPHIC

WASHINGTON, D.C.

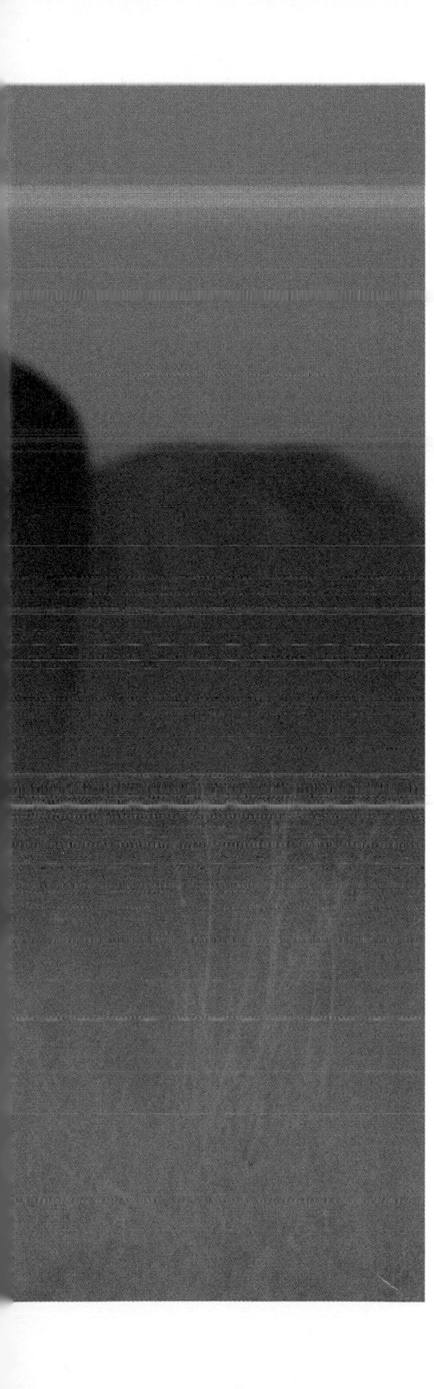

Contents

10 Foreword

12 An Ode to Duba

19 The Place: A Map of the Heart

37 The Eternal Dance

49 Lions: Shadows in the Dust

53 Tsaro Pride: Confronting Death

73 Skimmer Pride: Followers in Paradise

85 Pantry Pride: Stalkers in the Grass

89 The Hunt

99 The Chase

111 The Kill

149 And in the End

164 Making of the Film

172 What You Can Do to Help

174 Acknowledgments

Page 1: *A lattice of grass is the best camouflage for a lion. Focused, and dissolving in and out of reality like a ghost of the savanna, she stalks her quarry.* **Pages 2-3:** *Duba Plains is laid out at its finest, half flooded under a layer of winter floodwater that the lions must wade through daily to get to the buffalo they need to hunt. They are a team, with a single goal.* **Pages 4-5:** *Alone against the hordes, a single female is ready at dawn for the hunt.*

Duba at dawn is a unique place, where dust and mist mingle at the feet of the buffalo.

Ma di Tau, our lone lioness, searches for her mate after a night of mayhem. He is nowhere to be found.

Since I first came across Dereck and Beverly filming along the Linyanti River about 20 years ago, I have considered them to be pioneers in what they do: living and filming in the bush, immersing themselves fully in the experience, not because it feels good to them, but for all of us. They have always said that unless they could fully understand and experience the highs and the lows of a lifestyle like this, they would never be able to honestly portray it in their books and films. I think that this comes across in their work very strongly. From time to time we'd meet in different locations around Botswana and swap tales of our adventures, and when I traveled up to see them there was always something fascinating and new from them, a testament to the amount of time they physically spend with their subjects in the wild.

From our distant perspective wildlife and nature often seem to be one-dimensional, conforming to the rules of what science discovers about them. Only by spending extraordinary amounts of time working within nature can you see beyond the rules, and this is what the Jouberts do best. The wild places of the world are there as a reflection for us, something important, not to be abused, or simply for our entertainment. These precious jewels, of which Botswana has some of the best and most well looked after in the world, need to be protected for their own intrinsic value, not simply as a revenue source. The films and books of the Jouberts, like this one, do many things for us. They impart the preciousness of these jewels—places like Duba Plains—and lead us to a greater appreciation of the value of wild places. These projects make us think as much about ourselves as about the lions and buffalo or elephants that are so often the subject of their work.

We in Botswana believe in the careful protection of our resources. The low-volume tourism policy has worked extremely well for us, allowing the best economic benefit and the lowest impact on the environment. On a personal level, I have to admit to enjoying the sight of a huge male lion as he moves athletically across the grassland more than any in the bush. The symbolic beauty of a beast with so much latent power and enormous potential, the ultimate predator, in his natural environment, is hard to better. To see this animal living with such confidence of his ownership of an Africa we think of as our own is humbling. The humility of these moments is certainly not lost on Dereck and Beverly, as you can see in their words and photographs, a remarkable reflection of the place they love and have become so much a part of.

I personally hope that they never get comfortable, and that the hardship of getting stuck in the marshes and deserts of Botswana and living in a tent never loses its allure for them, because I look forward to many years of adventures together and discussions with them, and of course the next film or book.

Lieutenant General Seretse Khama Ian Khama
President of the Republic of Botswana

Used to the soft, wet land underfoot, marsh lions follow the buffalo anywhere and everywhere around the island. No place is inaccessible, no place too wet or too difficult for hunting.

An Ode to Duba

"This should be on the big screen!"
"I agree . . ."

THIS CONVERSATION WITH TIM KELLY, President of National Geographic Global Media, formed the basis of our agreement with National Geographic in 2005 to develop a feature film we would call "The Last Lions." We kicked around the creative expression of what that might be. In our opinion, a film about lions, without showing what these massive super-predators do—kill—is a view of the world that we know is naive and silly. For us, the greatest joy and happiness is in seeing things as they really are.

"A lion version of the movie *300* (a dramatic, against-all-odds account of the Spartan-Persian Battle of Thermopylae)" was discussed, and yet when we cut that version, we realized it seemed as fake as the idea of hiding every kill delicately behind a bush. There are many ways to tell the story of lions: through the eyes of the cubs, males, or females. But ultimately we got to the point where we looked around as a creative team and said, "You know what, let's just tell it the way it is."

The reason lions should be seen on the big screen, and what appealed to Tim early on, is the expanse, the sheer scale of Africa when you are standing on the edge of a savanna and looking out. It is wide, high, deep, and sweeping, and yet, as you are taking it all in, mesmerized, a tiny malachite kingfisher the length of your thumb flits by and seduces you further with a splash of color. It is a journey into a visual, auditory, and sensory realm that cannot be expressed completely in words or film. Yet we've always wanted to try.

This was our opportunity! We had completely fallen for Duba Plains some years ago, and had followed it during a rare moment in time—from inception! There are not many places in the world where you can say you have seen the place develop, or grow. But Duba is a small island in the Okavango Delta that has formed only within the last 20 years, thanks to the work of termites, annual floods, and the various rivers and channels that flow through the delta. Everything is new, and that excited us. It excited us in the same way it does when you stand in Hawaii and see lava spewing out from a volcano into the sea, forming new land; or when you stand in a great towering forest with a seedling sprouting up at your feet. These places are precious because they are innocent and untouched, and, in Duba's case, filled with lions!

Filming for the big screen is an exciting challenge because it is more than just television in a larger format. The structure of a film is different; the imagery has a stronger resonance. Going on to the big screen also meant that we could somehow engage with the audience in a way we couldn't on television. There is a deeper trust with a theater audience than with those at home who are having dinner or are distracted by their family. In some ways the virginal island of Duba is a little like the theater—uncluttered, not hyped, and with enough time and space to breathe. In this film we are able to have a one-on-one conversation with each member of the audience and, at the same time, be a simple window to allow that same conversation to happen between the audience and Duba itself; perhaps that has been our main goal throughout this project. The arrival and formation of the Tsaro pride on Duba was not straightforward. No lion story ever is. A young pride of lions made their way across the delta's waters to Duba. A female and some cubs, separated from a bigger pride for some unknown reason. Later arrived a mixed pride of seemingly unrelated females that had lost all their subadult cubs. Whether they were related to the lone female or not, we will never know. This story then is the genesis of the Tsaro pride, the start of lions (certainly of our generation) on Duba island. But within this story we

With two cubs left, Ma di Tau holds them close. She is a good mother in an impossible situation struggling to hunt alone with two cubs to feed.

found a much greater story, the story of all lions in that they gather; form new prides as subadults, mostly as offshoots from a matriarchal pride; have their own cubs; and stabilize. But then, in time, they seem to flourish, bloom, and wilt as a single unit after many generations. We've seen this twice before in our lives; whole prides die off with not a single survivor to carry on the genes in that territory. Some cubs may have grown and been scattered to the myriad of territories nearby, but this birth territory sees its pride collapse.

In many ways this story seems symbolic of all lions. When Beverly and I were born, in the mid-20th century, actual numbers of lions were sketchy at best, but some researchers placed them at around 450,000. Today, or at least by 2002, the best assembled and official International Union for Conservation of Nature (IUCN) number was 23,000. Our investigations indicate that lions have not been doing well since then, and even if we liberally project that curve to today, it is unlikely that the number of lions in the wild is more than 20,000. These animals are on the brink of collapse, and their decline, I am sad to say, has happened on our watch. It happened while we were filming and writing books on lions; while our generation got entangled in endless debates about whether hunting was good or bad for conservation, whether to allow elephant hunting or culling; and while we were wasting thousands of man-hours on talkfests about conservation. We carry the full burden of responsibility for the damaged environment that hangs in tatters around us today. Lions are just one of the casualties of our arrogance and lack of attention. In case I never get another chance, I apologize on everyone's behalf.

The film and this book, in some ways, are not a celebration of lions, nor are they a story of the original lions of Duba. The stories may not even really be about lions at all. They are serious appeals for us all to understand that if we continue down this destructive path, not only will we truly be talking about the last lions at some point, but we will be dealing with the last of everything. Fellow Explorers at National Geographic are reporting very similar results across different disciplines—deforestation, declining fish populations, coral reef bleaching. Lions are just one of the tips of this giant iceberg. These apex predators are so important to keeping whole ecosystems alive that with roughly 20,000 left we are in real danger of taking the engines out of the finely tuned systems that we depend on for so many things.

For example, Africa has a roughly $80 billion a year eco-tourism industry. National Parks rely on the revenue, as do millions of people. Our research shows lions are the major attraction. Most people would not go on safari if they would definitely not see a lion! Losing $80 billion a year would devastate Africa's economy.

Beyond the environmental and economic reasons to look after lions carefully are the deep ethical and spiritual importance they hold for many people. There are Zulu clans called the Ngonyama people (Lion People), there are statues and flags, etchings and drawings of lions around the world, and there is a warmth in knowing that lions still exist in wild places and that someone is caring for them. We hold their existence dear to us, as if we delicately cradle that wildness in our souls, something we'd hate to see disappear as we become more automated, more artificial, and let go more and more of our natural side.

We love them, we hate them, we admire them, and perhaps, we wish we were more like them. We know that our souls, the very essence of what makes us human, would shrivel if they disappeared. *The Last Lions* is a film designed to make us think about that paradise where lions live, revealed from behind its green veil by taking a journey with them. In defining this story, we went back to our original footage to look at the birth of the island and the first lions we saw arrive and venture out across the new tributary onto the island. In viewing our time with the Lone Lioness, Ma di Tau, we realized that in fact her lessons were our lessons. Her struggle to survive was our struggle; her battles were our battles. Perhaps this is consistent with any study of the natural world anywhere.

Tsaro lionesses ghost through the dawn, heading toward the buffalo.
FOLLOWING PAGES: *The pride watches from afar as Ma di Tau hunts the herd of buffalo.*

The Place: A Map of the Heart

Duba Plains and the series of islands around Duba were formed, like many islands in the Okavango Delta, by the constant weave of the myriad channels and rivers; its annual whimsical flood; and the miraculous work of termites, who build their castles first as large as a vehicle, then a house, and finally, after years of breaking down and rebuilding and change in the acid content in the ground, an island. Floating papyrus, often broken away or discarded by hippos and elephants, ends up lodging against a small inlet and stopping. The water flows under the papyrus until it grows and often sends down roots. The channels change, and in time there is Duba, a group of Phoenix palm islands with a few dozen warthogs, 2,000 lechwe, 14 tsessebe, the same number of wildebeest, some kudu, baboons, a scattering of other new residents like aardwolves and hyenas, a couple of leopards, some elephant. And although that sounds like quite a lot, there are very few other local residents—no impala, giraffe, or zebra, for example. Oh, except the lions and the buffalo.

At Duba, the buffalo move in a set pattern that gets interrupted regularly and reestablished. The lions stick to their territories plus a small overlap, waiting for the herd to circle into their hunting domain. The buffalo use every trick they can to disappear and deflect attacks; each day is a race to minimize the damage, but almost every day there is damage. On average we record 15 kills a month. Most successful kills are preceded by multiple attacks and attempts that take the lions on hunts through the water and through the midday in long, relentless chases that eventually wear the buffalo down or wear the lions down. It is never certain who will be worn out first by the process, but each day this goes on and on and on. It is why we call their relationship one of relentless enemies.

The scene has been set. Those are the participants. I won't call them players, as we would use the word to refer to the actors in a theatrical scenario. The word "players" implies play and fun, which is completely the wrong sense of this drama. This is about death, whether our own or an animal's, and these are important moments for everything around, including the observer. Understanding more about the hunt and the kill, as well as our own feelings about life and death, is what this is about. Duba has become a special place, a spiritual place for us and for many who visit it.

Masters of the underwater in the same way as lions are of the open grasslands, crocodiles are also a symbol of Africa.
FOLLOWING PAGES: *Space rather than place gives these plains their life and their allure.*

Clear tannin-leached water does little to hide the hippos and crocodiles in the river to the north of the buffalo range.
OPPOSITE: *The Ngoga River slithers through the watery papyrus desert that forms the southern boundary of Duba's buffalo movement.*

Diary entry July 2003

Duba is a wonderland like no other. It has rolling plains like the Masai Mara, yet it is quiet, uncluttered by humans. It is a swamp of crystal clear water, dotted with forest and palms. Yes it has some . . . issues, mosquitoes get bad at times, vehicles get bogged down, lions won't hesitate to take whatever looks inviting, so for humans it may not be perfect, but that is what makes it perfect for us! We drove out today to film and got so badly stuck it took until midnight to get out. Leaches must have sucked a pint of blood from me by the looks of it and the July winter water was freezing. The first lesson to be learned is that we all make our own paradises.

The dialogue inside the vehicle sometimes goes like this: "Beverly, don't panic, we'll be fine . . . don't panic . . . OK, now is a good time to panic!"

All inhabitants here must learn to contend with the deep rivers that surround them, and all these rivers represent.

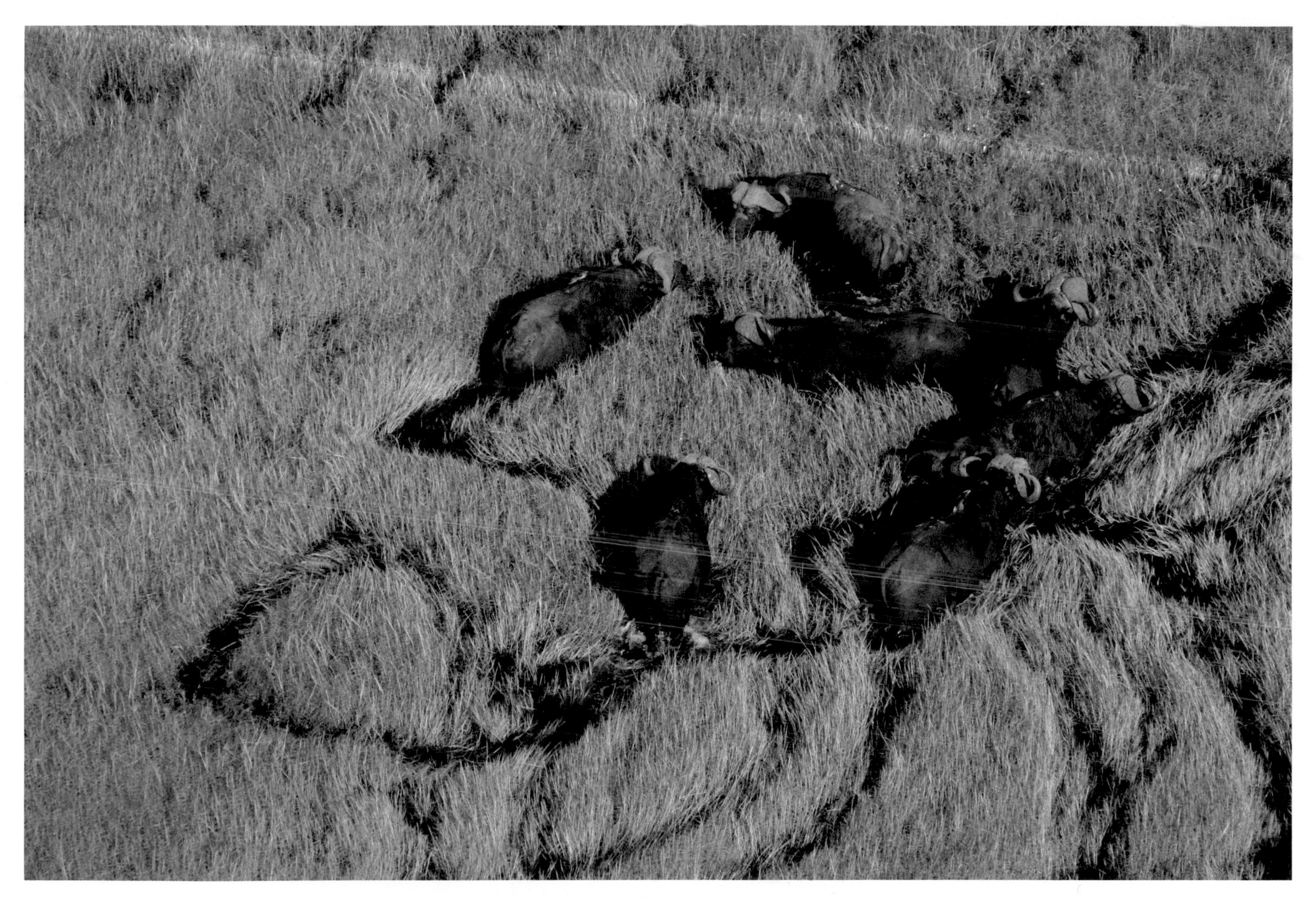

The old bull buffalo vacuum up the soft grasses and delicately avoid head-to-head confrontation at the same time.

The herd, more than a thousand strong, marches from one end of the island to the other, avoiding as best they can the ravages of the lions but staying within the boundaries of the ring of deep water that surrounds Duba. **Opposite:** *Hunting in water for hours each day has forged heavily muscled chests, necks, and upper legs, making the lions of the Tsaro pride the biggest we have seen.*

At the water crossings the lions force the buffalo herd to move faster, using the drag of water and the panic of confusion to select a victim.

WITHIN THIS ANCIENT RELATIONSHIP, a new, organized, and systematic trend among the lions has emerged. Lions are the great communicators of the bush, and communication is what holds all social structures together. Yet somehow, perhaps in the same way that two people find a common purpose just by being together for a long time, the lions of a pride seem to reach a level of knowingness without seeming to communicate visually or vocally. They move into a hunt as if they have a map in their hearts that each knows, a map of each hunt, a map of every move that they have traced either in their minds or by experience time and time again. Lionesses leave or approach the buffalo as one, silently, and circle around the herd to a position they seem to know, and into a pattern of a hunt they obviously understand. The strategy has worked before, and the more they slip into the pattern the more they succeed and the deeper the strategy is etched into their consciousness. And all of this happens without a sound from these ultimate communal hunters and social predators of the wild.

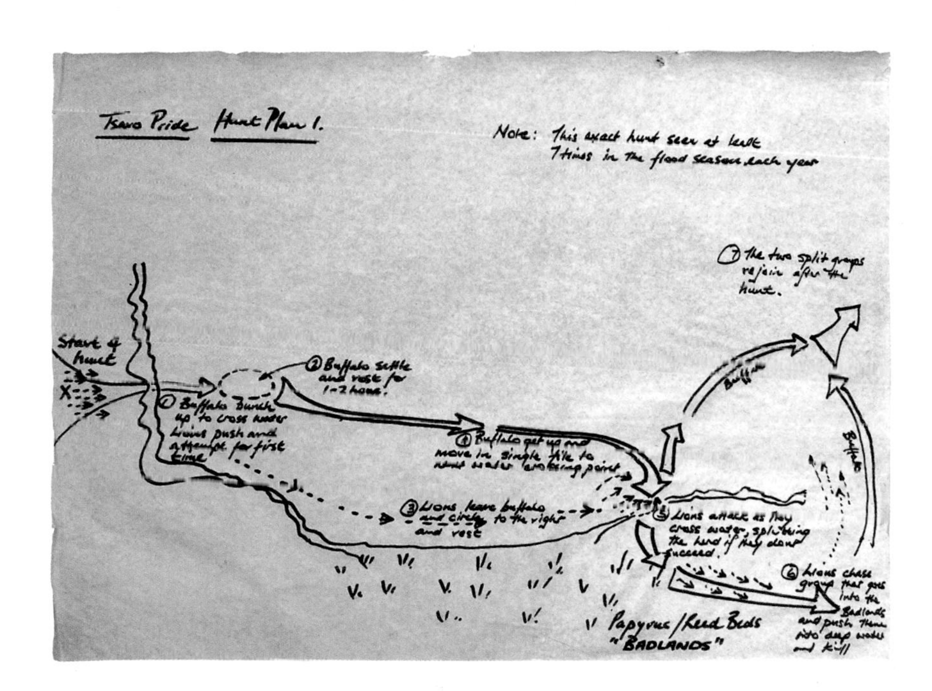

Swamp cats at work—endowed with heavy triceps, neck, and chest, they are propelled by thick sprinter's thighs and lithe lower legs.

The Eternal Dance

Time and space have forced these two animals together. Their dance is eternal, their destinies intertwined like one beast, a double-headed shape of amber and black, working together like muscles in the same body.

They dance again and again, each day, each to the same rhythm—shapes in the grass, closer and closer, an explosive snort of alarm, drumming legs, heads up and charging. The shapes duck away and disappear, the grass waves gently in the wind. Too many smells float around, confusing the direction of death. But even with the best plans and avoidance, the shapes hunt on.

In the end, someone dies. It is the way of this place, the way of Africa, where there are simply those who hunt and those who feed the hunters. It is lions and it is buffalo and it is Duba.

Wars are not fought for survival. They are fought for ethics or religion, ideologies, space, or resources. So to associate this relentless interaction between the lions and the buffalo in terms we can understand as war is erroneous. It is definitely one between predator and prey, and yet there is a beauty in its enactment. There is a playfulness in its execution that parallels the heroic battles of mortals in the Greek tragedies, as entertainment for the gods. Unlike the battles of Troy Homer tells of in the *Iliad,* this dance illustrates an interdependence as much as a conflict. Although the benefits seem one-sided, the buffalo thrive under the pressure, in a Darwinian sense.

This eternal dance is by definition "forever," although that assumes there will be dancers in eternity. As we stand back and watch this poetic shuffle, I cannot help but feel somehow sad at being left out. Throughout our own history we have danced our own dance with the great predatory cats, over millions of years, and it has taught us most about ourselves, more than have many of our other experiences. And while the biggest lesson we could learn was to survive, which we do amazingly well, it has extracted us efficiently from the dance itself. At our best, we place these big cats in cages or sanctuaries. At our worst, we wipe them out. David Quammen (author of *Monster of God*) suggests that within 150 years there will be no dangerous wild animals, like lions and crocodiles, simply because we won't allow them to exist. It certainly seems like that to me as I sit through endless debates about how best to accommodate our lust to kill them, or which populations of lions we can still save, or where our efforts are best spent. It sometimes feels like planning what furniture to save on the *Titanic,* instead of looking out for icebergs. And sometimes I close my eyes and remember the dance that still rages in my soul from four million years ago, and miss it.

So the eternal dance is theirs and ours.

The back-and-forth interplay of power between two of Africa's giants is eternal, harsh, and at the same time quite beautiful and essential.
FOLLOWING PAGES: *The dance is relentless. The buffalo attack the lions, which dive for safety, only to bounce back into the dance immediately.*

It seems as though it is not only individual buffalo darting out of the way that feels so much like a dance. The entire herd sweeping away in a set pattern ahead of the whole lion pride resembles a dance on a grander scale. These patterns so parallel the well-learned and rehearsed steps that make up a fluid dance that it certainly feels as though these lions and this herd of buffalo have been doing this forever.

The map below is another example of a hunt plan that the lions seemed to assume once the buffalo reached a certain point: breaking away from the more obvious tactic of following the herd, to sweep around into ambush, strategically, clinically, and using forethought. It has been said that the qualities that distinguish us from animals are the amazing abilities to be self-aware and grasp our past, present, and future. All animals, it was once thought, have a grasp of present, but past and future are beyond them. But these hunts, like so many behaviors, indicate that animals have a very well-developed sense of the past. In this case, they know that moving ahead to the water crossings has brought them success in the past. At the same time they clearly know about the future; again in this case the lions know that by going ahead they will (in the future) get into position to split the herd as it crosses the water, and increase their chances of success.

Each time we set a new level to distinguish ourselves from other animals, we find out something new about them, new information that makes us raise or change that bar. As we do, the lines between "us" and "them" blur. This process leads us quite quickly to the unsettling arena of thought about animal emotions and, even worse to many people, animal souls. There is little doubt in my mind that animals have emotions. It would be most strange for one ape to have developed emotions when others, in fact all other animals, have not. Emotions have armed us well for our development and survival. Clearly all animals have feelings like fear, a very useful emotion, so why not joy and even love?

These are cerebral as well as physical hunts, a blend of thought and action, a mind game resulting in a kill.

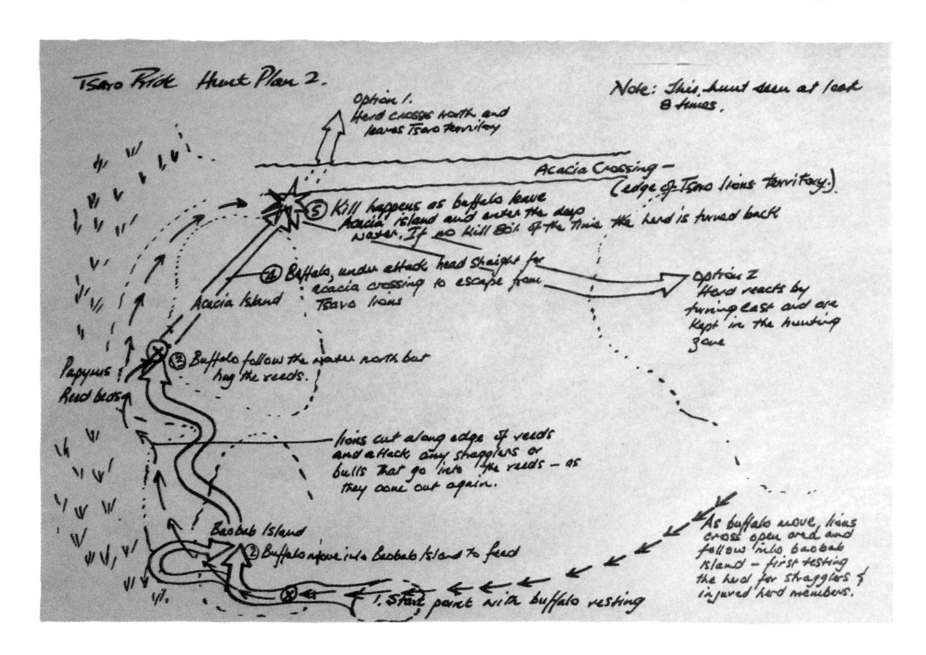

Darting for safety, running for their lives, dashing for cover: The lions in Duba are always at risk around the buffalo.

Diary entry January 12th, 2005

Up at 4am as usual. The best way to film is to catch the day napping and get out before the sun burns us to crisps. We find the female and cubs early, quite a lot of tension between them. It's just us, the lions and Beverly's birthday. Ma di Tau followed the buffalo. Water got deeper and we found ourselves driving through bad-smelling, buffalo-churned-up mess of mud and dung. Holding our breath, not against the stink, but because the truck was barely making it. Well until it stopped. Invested 4 hours of Beverly's birthday swimming in crap, and digging under the vehicle. Since we lost the lioness and the opportunity to film we enjoyed our lunch (mosquito free at last!) as the muck dried on our legs and arms into a sticky, smelly, contracting cake. I love this place.

Ma di Tau's cubs are a ball of energy. If they can't find high ground to conquer, they use her as that high point to battle over. The young male always seems to need to be king of the hill; his sister is less attached to material things.

Both bull buffalo and cows show their disdain and displeasure at being followed. Elaborate defenses, irritation, or avoidance tactics go on for hours some days.

The buffalo break out of the herd and defend aggressively, then quickly dart back before they can be isolated.

Lions: Shadows in the Dust

From our earliest memories of lions, we have held them up as icons of ourselves at our best moments: noble, brave, muscled, and social. Their biceps and triceps resemble ours, and their group size and structure is similar to what paleoanthropologists believe was our ideal group size in our first forays into the savannas as early humans.

We have erected statues to their image and glory, and we call our great kings and leaders by their name. In Botswana the president is referred to as Tau e Tona (the Great Lion). Richard the Lionhearted speaks for itself.

Similarly, another modern icon is the large bull buffalo, often considered the ultimate herbivore. As with the lion, we respect this animal in part because of its ferocity, its ability to kill people. We seem to idolize the very things that we've spent millions of years avoiding being killed by. It is the "noble enemy" relationship, which emerges from within us all from time to time. In addition, this strange evaluation of the buffalo is a hangover from an old hunting era, because it is said that they are difficult to kill. It can be tough to hunt them, I imagine, if you are armed with only a bow and arrow or stone tool. Today's rifles demystify the hunt; shooting a buffalo is quite easy now. But when a bull buffalo in its prime snorts at you and comes crashing through the palms at close quarters, slashing and smashing fronds as it comes, it most certainly gets your attention.

If you ever get a chance while out on safari one day, try to lift a set of buffalo horns over your head. Your arms will strain with the effort. These buffalo wear these horns and wield them with the dexterity of a fencer using his foil, but on necks built not only to hold up those horns all the time, but also to take the battering impact of an opposing male, who will drop his head and charge with the speed and intensity to knock his opponent senseless and off his feet.

Sometimes in these battles with lions, a bull will drop his horns and present a gnarled beaten boss, that solid dark battering ram that protects the skull, and charge. The lionesses know that it would be a bad mistake to take this threat on directly. Images of a puffing steam engine often rise to my imagination as the buffalo kick up dust in a misty sunrise and charge at the lions.

These two opponents are locked in a relentless interaction, here in this place and in this time. But so much can change. What happens in Duba Plains, this unique set of islands in the Okavango, will be different from what happens in East Africa, or southeast of here in South Africa. We have known some areas where the lions balked at hunting anything as dangerous as a buffalo.

This is not the case on the plains of Duba. All animals may have a baseline of characteristic behavioral patterns, but they also have a large layer of cultural, time- and place-related characteristics. Some lions just learn how to hunt elephants and then specialize in doing so. The Duba lions, as bold as they are with buffalo hunting, have never killed an elephant. Lions seem to test and learn in much the same way we do. We are best at finding patterns, fitting those patterns to our own past experiences, and learning from them. Here, the lions have learned what works for them. In time something will change, and they will learn something else that works better for them. To lock them in permanently as being daylight buffalo hunters, as they are now, would be a mistake.

Masters of all at Duba, these males are probably brothers, age 13 years.

The males in the pride seldom hunt or even participate, but they always dominate the first feed.

Tsaro Pride: Confronting Death

I used to understand what a pride of lions was, but more and more we discover that a pride can be comprised of lions with different relationships. Our film shows a mixed pride—a conglomeration of potentially related females. I strongly suspect that what we know today as the Tsaro pride was formed out of the need to successfully hunt the massive buffalo herd. I also now suspect that some of the characters we know well and that have become part of our lives came from different backgrounds.

Part and parcel of the Tsaro pride formation, however, is this gathering of lions in a fairly dysfunctional relationship, where lionesses—notably two, an old female and Silver Eye—kill cubs. To the best of our knowledge, the Tsaro pride has killed 96 of their own cubs since 2005. I've mulled over why and what the ecological answer may be and have ended up with a simple answer. We share so much with animals—the ability to communicate; to have emotions; to think in the past, present, and future—why would we hold exclusive rights to having screwed up families from time to time? Bad motherhood is certainly one possible cause of the low survival rate of their cubs. We have seen females following the buffalo for miles, marching on with little concern as the cubs fall farther behind.

One day, out of the corner of my eye, I saw a lioness lean back tensely. Suddenly she stood and pounced into the grass, raked her claws through the grass, then dipped her head down, teeth bared. This took just a few seconds, but in that time she flicked one of the four-month-old cubs at least three meters into the air. The cub was dead when it landed, guts spilled. The other lionesses scattered, some away from the scene, some toward it. The other cubs ran off across the plain. We noted the lioness's ear notches. She was cubless.

The pride was in disarray for hours, with the lionesses calling softly and approaching one another cautiously. Then in the afternoon one lioness approached and sniffed the cub, called and sat down, then slowly stripped off the skin and ate the cub. Stunned, we took our pictures silently and then backed away as if we'd seen something private and illegal. When she stood and walked away, her suckled nipples showed her to be the mother.

The cub killer continued her aggression. Twice attentive mothers stopped her attacks, but it was not enough. Five cubs disappeared during the night that week. Obviously affected, the pride followed the buffalo but did not make a kill for the whole week.

Soon, a small young female with clean ears (no cuts or nicks) decided that it was time to introduce her two new cubs to the pride. We could see them coming across the grassland, and so did the cub-killer!

She walked like a lioness going to greet her companion. Then she arced around the female and started trotting toward the cubs, ears forward, neck arched. This was the cubs' first encounter with any lion besides their mother, and it was bearing down on them fast. The mother had been outmaneuvered, and the cub-killer (possibly her own sister) was between her and her cubs. But in an instant she was right there with so ferocious an attack that the cub-killer collapsed in submission and balled up against the barrage of blows. The cub-killer turned on her back and submitted, stopping the attack, and crawled away on her belly to a safe distance. The cubs watched with big eyes, secure for now.

We named the female attacker Silver Eye. Her eye had been gouged out and only a silvery skin remained. We had observed her for several seasons as she tried to dominate the Lone Lioness, and she had become more aggressive over the years. She may have been the leader of the cub-killing trend, potentially related to her eye injury, which put her at some disadvantage. Originally, we felt Silver Eye was attacking the Lone Lioness's cubs because she had been responsible for Silver Eye's wound, but we don't think that now. It is too extensive a retaliation. It might be a natural reaction to the confines of a small island, and perhaps the lions have no other way of controlling their numbers, but it is a harsh method of keeping stability.

The Tsaro pride, two huge lions and nine huntresses, dominate the heart of Duba.

Silver Eye, as we called her, rather than the Blind-eyed Female, took on a certain beauty once we'd named her. Lions are supposedly all of equal rank, but Silver Eye stood out for more than just her features. She was an excellent hunter, always taking risks either as the first female to attack or among the front line. **OPPOSITE:** *The other large-bodied Tsaro females carry the typical remnant stripes on their backs. We'd seen this before in Savute, but it isn't evident in all lions. We suspect it's a ghost stripe from some past time when lions had stripes.*

The small cubs took fright, but their mother intercepted and attacked the female, which clearly intended to harm them.

One day the tension between the mother and her cubless sister exploded into a full lion fight when the sister ran toward some young cubs about to be introduced to the pride for the first time.

This island wouldn't have much left on it if every cub born survived. Ultimately, just as lions have a lifespan, so do prides of lions. We have seen whole prides simply die out. This happens when more lionesses die than cubs are incorporated into the prides. What we have at Duba with the Tsaro females is a pride that may have a traditional (or ecological) ideal size of nine females. There is no room for more lions in the pride; the excess will die or move on. Right now they are dying young.

Nevertheless, the Tsaro pride is flourishing. These are the largest lionesses I have ever seen. Thick necks and bulky bodies distinguish them from the other prides nearby. The reason may be their almost exclusively buffalo diet, but it may also be the exercise: constant following and then big attacks on large prey, which often need wrestling to the ground. Either way, these lions are huge.

Their special hunting technique is different from that of the other prides, and I call these lions "The Confronters." They seldom stalk, but walk in boldly. They usually start the day by walking up to the buffalo herd to get them on the run. Then they scan for any weaknesses, as if paging through the telephone book, looking quickly for the telltale sign of a limping animal, young or old—preferably young. The buffalo run, but then return to chase the lions. The lions hold their ground, perhaps as a test to see just how far the buffalo will push and be pushed. At some point they may chase in directly to get the herd running, then follow through with an attack. This is confrontational and relentless hunting. Some days the hunt lasts six hours. Recently the lions have had success in attacking from the side, splitting the herd and waiting for the stragglers or a smaller herd anxious to join up with the main herd.

Tsaro pride females seldom look at the herd as it moves forward after the initial morning testing, and it is strange to watch them all turned away from a thousand buffalo. But they are looking for stragglers, and even the splinter groups of bulls that lag or weave off to the flank of the herd draw their attention. They seem to know that at some point these splinters will try to rejoin, and when they do, if the lionesses can get between them and the herd, the hunt is on. Buffalo always seem weaker as a splinter group and try to join the herd, often in panic. When the herd is nearby you can predict the route this offshoot herd will take to get back to the main herd. So can the lions.

At a certain time of the year these lionesses hunt calves almost exclusively. As the birthing season starts the lionesses change tactics. They push the herd until the calves drop back, then they run in and collapse the calf just long enough to do some damage before the herd returns to rescue it. If it is a long hunt, the herd may leave, and only one kinship group stays behind to defend the calf. At that point the lions want to get rid of the rest of the buffalo so they can eat their catch in peace.

Many of the Tsaro pride's hunts, however, have ended in two or more kills: Lionesses see that a mother may be vulnerable as well, and give chase again even though they have one buffalo down already. For some reason female buffalo collapse almost instantly. An animal that size should be able to put up a stronger fight. Even impala battle their attackers harder, in my experience. Male buffalo stand and fight for up to an hour, yet a female that is only 20 percent smaller than a male falls in seconds.

As a result of many lost cubs, the Tsaro females breed out of synchrony at the moment, so larger cubs and small ones mix in together, making it tough on the smaller ones and making new mothers more defensive, often with fatal consequences for the older bullying cubs.

In recent years, fewer Tsaro cubs have survived into adulthood. The long water crossings are difficult and dangerous. Crocodiles, drowning, and simply falling behind may be the causes in some cases.

Cubs in the Okavango have to adapt to water very quickly and conquer their innate dislike of getting wet. It is a difficult lesson for a cat, but even from their first days the Duba lions take to water easily.

To watch in the early morning light as the nine lionesses, almost identical females, fan out and start to move in on the herd, is enough to take one's breath away. The sheer beauty of this moment always gives me cause to breathe in and fill my lungs with life. It feels like the moment before some great event, a storm perhaps, a reading by the Dalai Lama, or a birth. Its purity is of a type that is so clean that you want to capture everything about it. You want to freeze the moment in a photograph, symbolize it in a painting, detail it in slow-motion film footage and sound: a crystal so perfect it needs its own showcase. It is that moment often reached for in stories and films, when out of chaos something suddenly synchronizes and is whole; the moment when a symphony, Mahler perhaps, rambles and bumps and then quietly finds its harmony and becomes one, not an ensemble of players, not nine lionesses, but one melody. And you know it isn't folly. Someone is about to die.

I don't get excited, nor do I get as anxious as others do when they watch the start of a hunt. But something fills my body with the profound—and everything seems to sharpen. Perhaps it's my need to be professional and take observations, as well as get into the right position without interfering, and then film it all. So we are busy, but the sound of grass brushing against their faces flicks in my ears, the crunch of their feet is agonizingly loud. An alarm call by a francolin indicates that the male is coming in (and potentially blowing the hunt). The wind drops and flutters a piece of hair into my face; it has changed direction and the buffalo will pick up their noses soon. They do, and they flick their tongues into their noses to add moisture and enhance their sense of smell. They know, as we do, that the time has come.

After that it is all mechanics, the job of getting it right, for both the lions and us, the magical moment over but still burning inside.

The Tsaro pride have learned a few other things.

The buffalo use the whole of this island and cross over to it in three places. But the Tsaro pride live only on this island, so the trick is to keep the buffalo in their territory. Some would say that forward thought like this is beyond animals and is a purely human trait. Tsaro think ahead and turn the herd back as they head for the river if they are going north. Once in a while the lions' hunger drives them to attack as the buffalo are crossing, and they lose the herd for days at a time, as the buffalo swim toward an area called Paradise and an adjoining lion territory. But most of the time they ambush the herd just before it crosses the river, effectively turning it back home.

On the other sides of the island the lions know that they can push the herd as much as they like, because in the south an impenetrable bank of reeds and papyrus will eventually stop the herd and send it bouncing back—into the waiting lions.

To the west, in an area we call the Badlands (because it is bad: broken springs and bent chassis are the reward for attempting a crossing for us), the buffalo struggle over uneven ground and risk a strained leg or other injury. The lions will go in after them, pushing the herd to water and through difficult terrain, and then sweep for the injured.

To the east, the herd can be followed, but there the lionesses always spend their time aggressively looking around for the rival Pantry pride. (These are not names given by us, by the way. These lion prides have been named by the guides at Duba Plains camp, the only safari operators in this area.)

Soft mouths for carrying cubs are also hard mouths for killing buffalo.

magic and contemplation, heated discussions about Nietzsche or Descartes or sometimes hours of wordlessness punctuated by panic and chaos as lions get up and run and attack or something else wonderful happens. Today was not one of those days. We sat in the sun avoiding the chilling wind semiwordlessly, and by 3pm decided to visit the lioness. As we drove over to check on the lioness, Beverly noticed from her high perch on the roof that she was with a companion, a dead buffalo calf that she had obviously killed silently while we (and the pride) weren't looking. There goes that day and any hope of footage!

Male lions truly own Africa, until someone else comes along to challenge for territory, but for now, the soft grass, the females, and even the stormy sky above belong to him.

The story of Duba and its lions cannot be told clearly without talking about the males. In our film, we elected to follow one lioness, the Lone Lioness, or Ma di Tau, and her journey onto the island. The males that now come onto the island tell a stunning story of their own, one that is mirrored more and more in Africa.

Between the first insurgence on to the island and today, two males found their way in. They ruled for some 14 years as the "Duba Males." They were very old, both with reasonable manes—one darker, one lighter—and faces like an old carpet, bitten and beaten from years of fighting. One day, while resting in bushes, they were cornered by buffalo and attacked. Both died. One was killed and mauled on the spot. Now, a male called the Skimmer male has moved in and dominates the pride. Ironically, he has no male challengers, and while he is still young, he has been around for three years, close to an average male lion's tenure elsewhere.

If you consider the population numbers—20,000 lions—we may have only 4,500 male lions, yet the United States still officially allows and sanctions hunting of those lions at a rate of 500-600 a year. The U.S. imports 556 lion skins as trophies each year from safari hunters. If this is not totally insane and unsustainable it is so close that it will drive collapse faster. Hunters say they only take old, excess males, but the Duba situation confuses that statement: The Duba males were breeding up until the day the buffalo killed them. Also, when males grow old and are ousted from their territory, they lose their manes and are no longer attractive hunting trophies.

Much has been written about a lion's mane. It is a burden in a hot climate. Some suggest that a black mane makes the male at least 12 degrees hotter. The mane is unlikely to have evolved as protection against attacks, since females fight almost as often as males do. It is a very visual signal that often works against the whole pride when the male wanders along to meet up with his hunting females and scares off the prey. It won't prevent a fight; if the challenger wants to fight for territory he will, but the mane certainly signals that a dominant male is in residence and prevents a challenger from thinking the coast is clear.

Prior to 1995, this area and adjoining ones were trophy-hunting areas. The shooting of one male can result in up to 30 lions dying because of the structure of the pride; the coalitions that males form, especially with a mate; and infanticide by new males. Trophy hunters seriously damaged the lion populations in the northeast where we lived, and we watched the collapse. I don't know what level of hunting happened at Duba; however, with each hunter needing at least one male lion from this area every 21 days for the hunting season, I can imagine: Every male was shot, prides disrupted, challengers of young age shot as well, until the pool was dry.

We watched some males one day as they surveyed their vast domain from a termite mound the size of a car. They were old, and we'd known them for five years already and seen them most days. They must have been too small to be shot in 1995, perhaps three years old. However, something happened one day while we were sitting with the pride to indicate that they were at least aware of what was going on around them as young cubs.

At the camp, at least three to five kilometers away, the manager accidentally set off a "bear banger," a loud gunshot-like explosion. The lionesses, younger than the males, didn't flinch, but the two males leaped up and stared in the direction of the shot. They slunk off into the thickest thorn bushes Duba provides, looking alternately at us and back in the direction of the camp. They remembered!

I know that many people feel the need to shoot a lion to prove something, and even more people are apathetic toward hunting. I have some difficulty being apathetic. I have an even harder time liking those who need to shoot lions. I don't like people who need to show off their prowess or their egos through trophy hunting. I think that our society deserves better. We need men and women who can lead us into a new era that offends nature less, takes from it less often, and thinks less selfishly. Thankfully, that time is coming.

Against a constant backdrop of buffalo, the lions of Duba are highlights on a dark canvas.

Skimmer Pride: Followers in Paradise

When we started our work in Duba, and the lions started moving across the river to the island, there were no residents there. After a year and a half we started seeing the Skimmer pride sneak across, which they did with a comfort and willingness to cross water that is unexpected of lions. In contrast to the Tsaro females, which are so extraordinarily large, these females were more normal in size. Although adjoining prides are often related in isolated areas like this, these lions just have not developed the bulk of the Tsaro females. But the Skimmer pride live to the north, outside of Duba, and make only occasional forays onto the island if they are desperate.

It was a drizzly day in summer when we found them hunting. The buffalo had escaped from the Tsaro territory, with only one casualty, and had crossed through the water, leaving the nine lionesses chest-deep in water watching their prey disappear.

Earlier they had killed a calf in the water in spectacular fashion, spraying water everywhere and hauling the baby away from its mother's side. The herd moved north. Having been stuck up to our axles the day before in water we call the Smelly Crossing, for obvious reasons, we ventured around through the edge of the hippo pools and got ahead of the herd—and found the Skimmer pride.

A lone female buffalo kept turning back and calling. She must have been the mother of the calf that had earlier been killed by the Tsaro pride. The whole Skimmer pride woke up and started to run in.

They stopped and watched the herd go by, waiting for the stragglers, and then finally went into their usual strategy: they followed. I call them "The Followers." Their special technique is to follow each move the buffalo make, just a few hundred meters behind, and to stay out of their way. Sometimes the buffalo turn back and confront them, and the whole Skimmer pride then heads for the trees, scrambling up to safety. However, it is unusual to see the horizon dotted with lions swaying uncomfortably in trees.

The pride eventually spotted a limping adult female. The four lionesses went into position and ambushed her. After days and months of driving through meter-deep water, our filming truck had long since given up any semblance of a braking system. As we drove up to the two females wrestling the cow to the ground, I couldn't get the car to stop, so both of us grabbed cameras and let the vehicle shudder to its own halt some way off. We filmed and photographed the kill from the ground, which elicited only the slightest glance from one of the females, now intent on getting her prize off its feet and onto its back. She succeeded and we backed off. The incident went off without a problem as the rest of the lionesses and the cubs raced in and killed the female buffalo. It was a typical Skimmer hunt.

The interactions between the Skimmer pride and the buffalo take on a certain playful feel mainly because of the cubs, with buffalo snorting and chasing and cubs ducking and diving through the grass and doubling back to pick up the game as soon as the buffalo run by. The lionesses watch, doing their dual job of scanning the herd and keeping an eye open for cubs getting into trouble. These are good mothers, and they have reared cubs successfully in the past few years. I suspect that the pride will take on the female cubs and grow to eight or nine. The male cubs will be cast out to wander until they can find a territory of their own. At the moment there are five of these young males, and if they all make it they will be a formidable coalition. Maybe they will return one day to finally oust the Duba males.

Skimmer pride, in contrast to Tsaro, are successful cub raisers for now, rearing nine of their ten cubs.

ONE EVENING WE WATCHED as the Skimmer females, as usual, followed behind the buffalo across the water. The evening light was just beginning to fade and we had half an hour left before dark. Suddenly one lioness looked over her shoulder and saw that behind her a group of about 70 buffalo had been left behind and were anxious to join up. She turned and stalked back, wading into the water toward the buffalo, which were wading into the water toward her. She kept going, a single amber figure walking boldly into an approaching wall of black shapes. Then, when she was about 20 paces from them, the herd started to run—toward her. I thought I saw a moment of hesitation as her neck stiffened, and I was sure that retreat was definitely on her mind. She quickly conquered that thought, and launched into a head-on running attack. The herd turned, and in the chaos this one lioness leaped on and attacked an adult female buffalo, bringing her down in a bundle in the water. It was one of the most confrontational kills we'd seen. Emboldened by this, the rest of the pride ran almost straight past the female—still struggling to kill the cow—and in the distance killed another buffalo in the last light of the day.

It is quite rare for these lions to kill at night or even in the last light of the day, and there has been much speculation on exactly why. It was a puzzle that I was determined to solve. I asked scientists and lion experts as well as bush people whenever I could. In fact, I asked everyone. The puzzle is this: Lions have relatively small hearts for their body size, making them less efficient in the heat. Duba, like most places in Botswana, can get very hot, certainly not less so than Selinda and Savute, where lions are mostly active at night. Why should these lions hunt during the day?

An associate from the University of Oxford put it quite simply by saying that for a behavior to occur it should be a) a distinct advantage; b) the opposite of a distinct disadvantage; or c) possibly random.

Let's say an advantage might be that the lions can see their prey better in daylight (not true, their night vision is good, and besides, buffalo are huge black shapes against a light yellow grassland; even we can see buffalo most nights). Or, let's say that they can navigate the water better, with less crocodile activity (it's unlikely, although I'm sure that crocodiles may be more of a threat in the dark, but large tracts of this land aren't underwater, especially in the summer; the actual waterlogged area of Duba may be 20 percent at best, not enough to change their entire hunting pattern).

I can't think of other advantages of daylight over night for them.

Let's say the disadvantages at night forcing them to hunt during the day may be increased hyena activity at night (we haven't seen this, and have seen hyena takeovers during the day, so it isn't a determining factor). Increased competition from male lion takeover at night? The cooler nights slightly dissolve the daytime disadvantage to male lions. They are slightly less well adapted to heat because of their heavy and dark manes, which, in some cases, increase their body temperature. But on Duba, males are not an influence now.

Might there be increased nomadic male action at night to cause the switch to daylight hunting? (There are no nomadic males.)

So, all we can come up with is the third factor: that they do it because for some random reason they have learned to and it works. It really is a mystery. It certainly is not (as some scientist seriously suggested to me) that the lions are so hot during the day that they get uncomfortable sleeping and get up and go hunting. These lions sleep very well, no matter what the temperature is.

I believe that the lions here hunt through the heat of the day because that is the pattern they have become accustomed to, against all odds. I wonder when they might find a new pattern.

What we have now is almost a reversal back to what we found on Duba in the very early days when the island was forming. The Skimmer pride no longer visit. The Lone Lioness is often alone but hunts with the Tsaro pride, and there are no new males. And yet the buffalo increase in number each year and one pride dominates, possibly because they know exactly how to hunt them.

Often the Skimmer pride poach into Tsaro territory, swimming in at night and snatching a buffalo before the Tsaro females are aware of their presence.

Ironically, the young males of the Skimmer pride may be the best hope for the future of the Tsaro pride, even though they are enemies right now.

A prime example of the fluidity of Africa's top predator in motion. While Beverly snapped off a dozen masterpieces, I was cursing and slapping my camera around to arouse it from a mysterious slumber. The camera finally came on as the buffalo collapsed in the mud, and I had filmed absolutely nothing of the event.

Skimmer females, not quite as heavyset as Tsaro's females, still manage to live off buffalo and are skillful buffalo killers.
FOLLOWING PAGES: *When the Skimmer pride goes on the march across the marshy wetland, their pale bodies reflect the moonlight like pearls in the darkness.*

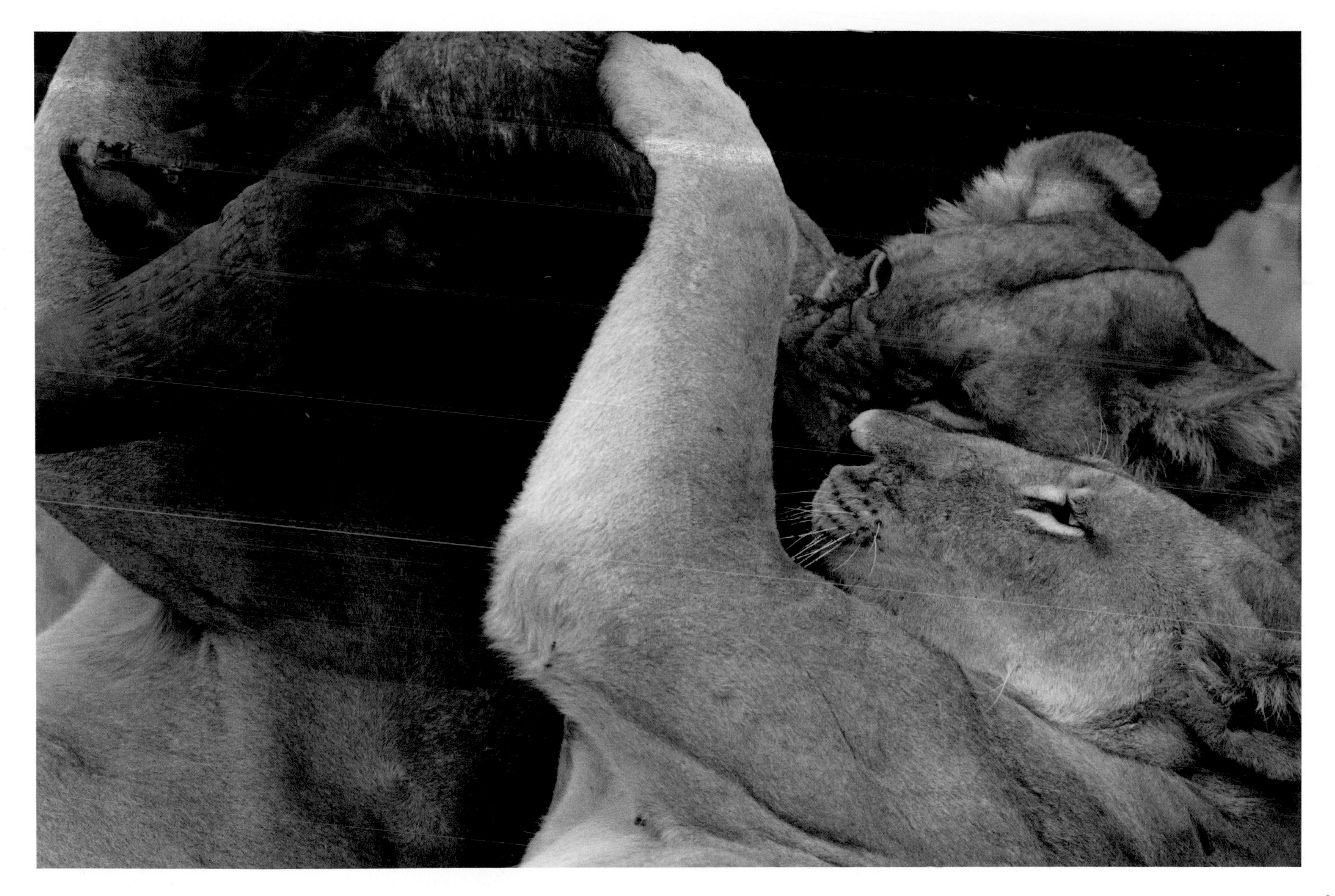

Pantry Pride: Stalkers in the Grass

The Pantry pride once numbered eight lionesses strong; one of them was quite fierce, possibly an older lion from the hunting days. Every time we saw her she made a point of walking straight up to the car and bursting into an angry charge. Once when we had a friend, Martha, with us, this lioness burst into a charge very close to Martha's side of the vehicle, with a roar so loud that it made the vehicle's sides rattle (and I am sure a few body parts within the vehicle, too). That same lioness once wandered into the camp and attacked someone walking at night.

One day the fierce lioness took on a buffalo that was just too much of a match for her, and she was sliced open across her stomach and condemned to a week of suffering before finally succumbing to death. One by one the rest of the pride fell to the buffalo or just disappeared. This pride had seen its natural life span, comparable to that of a lion, and for a while only three remained. They would still roar their defiance to destiny from behind our camp almost every night, and occasionally swim across the river to find the buffalo.

These lions are "The Stalkers," though, and their method of hunting buffalo, probably because of their smaller group size and maybe because of the toll the buffalo have taken on them, is to sneak in around the buffalo. They hide in the grass and finally explode in surprise, spooking the herd, and then, with darting eyes, find their opportunity.

One day one of the remaining Pantry pride lionesses caught a buffalo by the throat. She hung on during almost an hour of combat while the others alternately jumped on the buffalo's back and dodged the relentless attacks of a handful of supporting bulls that were determined to save their companion. One male put his head down and charged, hitting the lioness fully broadside over and over in an attack that would have killed most lions, and definitely would have at least made them release their stranglehold. But she took the hits and tensed her body against each successive blow. Eventually, without releasing her hold, she balled herself up under her victim's neck; the attacking bull was hitting his companion as its head offered some shelter to the Pantry female. Finally she collapsed the buffalo, and as its calls faded away the rescuers left and the lioness moved off to the shade. It was the last we saw of her—a gallant battle, won but ultimately lost.

That was the death of the Pantry pride. The remainder of the pride withered away and disappeared. Duba is a tough place where only the strongest survive.

The Pantry pride was once as large in number but not in physical size as the Tsaro pride. But buffalo attacks gone wrong have slowly whittled their numbers down to just three.

The young male of the Pantry pride will be chased mercilessly by the buffalo—as well as by other lions around Duba—until he forges a coalition with another young male and makes a pride his own.

Pantry pride are now stalkers, mainly because they lack the numbers for a frontal attack and are easily chased off by the buffalo if detected.

The Hunt

Since this relentless daylight hunting is so visible, the efficiency of the prides is misleading. Because they seem to be super-efficient, we tried to evaluate just how successful they really are. In judging this we had to consider what really constituted a hunt. Eventually we decided on the basic parameters: from when they arise from a resting period; go on a stalk, chase and hunt; and then declare it a miss (or kill) and go back to a resting period.

We had to define this because so often kills start with following (usually this would be a Skimmer pride tactic), or walking right in (a Tsaro pride technique). The buffalo herd stands or gathers together and chases or retaliates, and then finally runs. At this point the lions may attempt a kill and fail, but they carry on hunting again and again without stopping. This, by our definition, is just one hunt, even though they may try to catch three or four animals within that one piece of action. It isn't unusual lion behavior, by standards and observation throughout Africa, to run at an animal and get rebuffed a few times before either calling it off or succeeding, whether the prey is impala, buffalo, or elephant. However, by this parameter we calculated that in fact the lions at Duba have almost exactly the same success rate as lions elsewhere in Africa—somewhere around 25 percent. The difference, maybe, is that with these lions the length of the hunt is extended; they hunt and hunt and hunt, over and over, until they succeed. In other places in Africa the lions can switch prey after a few failed attempts, but not on this island. Here it is one predator and one prey, day after day.

But these hunts are most interesting to watch because the setups are so intricate, so tactical, and so varied, ranging from the bold to the sneaky. These lions display amazing tactics to outsmart the buffalo, testing their mental abilities as well as their physical ones, and that junction is where the best distinction between hunt and kill is to be found. It seems to me that the hunt is a mental effort, and the kill is mostly physical. That is why the hunt is more interesting. Not only does it take planning and knowledge of the field, confidence and knowledge of your own abilities, but it also takes forward thought.

Very often, as the hunt begins, we might see a lioness move off in a strange direction and we say to each other, "What is she thinking? What opportunity has she seen?" It is a chess game that requires moves to be thought out well in advance. And in the same way as a student of the game might see a chess move and silently say, "Ah, the Karpov 1971 opening," we might say, "Ah, the old Python Island water-crossing move." This is the hunt, and this is what lions are born to do: a special combination of thought and power as well as cooperation within the group. A lioness looks up and sees that another lioness has the gap between the palms covered, and then she moves into a different position, one that may be more exposed and thus spook the herd toward that gap in the palms. Sometimes individuals set themselves up to be the first attacker; on other hunts they may be the herders. Each hunt is a running, shifting battle that is unpredictable once it gets going. The buffalo are smart and are desperate to survive. The lions' job is to cut off that escape, to outsmart them, to force them into a position of weakness, and all the time ready themselves for the eventuality of a single buffalo breaking away and the attack it may need right then, in a split second—or the equal and opposite reaction to the sudden switchback of an angry bull.

Yes, this is the hunt!

The tactic: to split the herd and isolate the old bulls, then trip or anchor the massive bull, and finally to attack and weigh it down by sheer force.

Eyes watch from the distance, selecting the prey, calculating the risk, and weighing the reward.

The hunt begins with the intention. It doesn't end until the kill.
FOLLOWING PAGES: *The hunt goes through water, following the prey. It halts momentarily, until the next attempt.*

The counterattack can only slow down the relentless hunt and its inevitable finale.
Today, tomorrow, or the next day, there will be a conclusion.

The Chase

When the Tsaro pride approach the buffalo herd, the encounter begins with pushing. But then something suddenly clicks. You can see it happen. One day we got to the herd and found the lions all snoozing, no more than three meters away from the sleeping buffalo. It was bizarre; they had almost become part of the herd. Now they were getting so used to each other that there was no pretext of hiding. The lions knew they would get a buffalo sometime, so they slept. I joked that, in time, the lions would sleep in among the herd, using their bodies as cushions. This really seemed extreme, however; the buffalo were all resting, many fast asleep, with heads flat, and most even facing away from the lions.

Eventually, two hours later, the buffalo moved and the lions stretched and walked in. Now the buffalo retaliated, chasing the lions away, but each time the lions bounced back. The buffalo bunched together to defend. This went on for six hours, and then, suddenly, as the buffalo started to charge and chase the lions, something changed.

The lions stood shoulder-to-shoulder and started to moan and growl. We have often seen them do this when they are defending against hyenas or against buffalo that are keeping them away from a kill they have made. However, this time the buffalo were keeping them from making a kill, and the lions had had enough. A lioness charged the buffalo. It wasn't an attempt to chase after them and catch one; it was a charge, with head up and neck back, a running posture that is a threat display (usually used within the species or against hyenas at best, and sometimes against humans). The buffalo also sensed the change, as if from a game to serious displeasure, and turned tail and ran off. Suddenly the lions were on the hunt; now they ran in with their ears forward, not back, necks extended forward. The buffalo charged off into the water. The lions simply ran in after them as if the water didn't exist. It was an exciting and dramatic moment: seven lionesses running, looking for opportunities, darting this way and that, all in the blue water. They were like sheepdogs herding these animals together, making them panic, taking stabs at their flanks to keep them bunched, and waiting for someone to fall over.

Someone did. A calf dropped down, but quickly found its feet. The instant that it took the calf to recover lost ground was too much. A lioness dived through the waves behind the herd and landed on the calf. Both disappeared into the water. As they bobbed up in the water, the cow saw her calf under the lion and splashed in to save it. Her hooves chopped at the water and all around the lioness's head. In a cascade of splashes and wild flashing hooves and horns, confusion veiled the lions and threw the herd into absolute panic. It wasn't until too late that the buffalo realized that there was yet another beige shape in among their legs, reaching for yet another of their young.

Through all this we moved behind in a chaotic zigzag, hopping as best we could from one slightly drier-looking island to the next in the swamp, pushing through whatever we saw the lions manage, assuming that our vehicle was at least up to getting through water that lions could wade through. At the deeper water, where they had to swim, we decided to abandon logic and risk crossing anyway. Strangely, we made it through the deepest and muddiest of the crossings and filmed both kills. When it is all over, and in the presence of bloodshed, we settle back and talk in hushed tones or just digest the moment.

When the lionesses stand in front of charging bulls, they are at risk of being trampled.

A wall of lions going into the hunt always spells excitement.
There is a moment when the lions seem to have decided that they will hunt,
and they will do it together—a chilling sight if you are their prey.

Water is difficult to move through quickly. When lionesses run at their prey, it takes all their strength and energy, the most likely reason for their enormous bulk.

Distractions dissolve during the chase. The lions are locked onto one objective, and need to be. A large bull could turn and retaliate, with fatal results.
Following pages: *Here is an intoxicating moment when a lion transforms from a standing but alert animal to a charging, hunting predator in action, a killer at the pinnacle of its evolutionary development, doing what it was born to.*

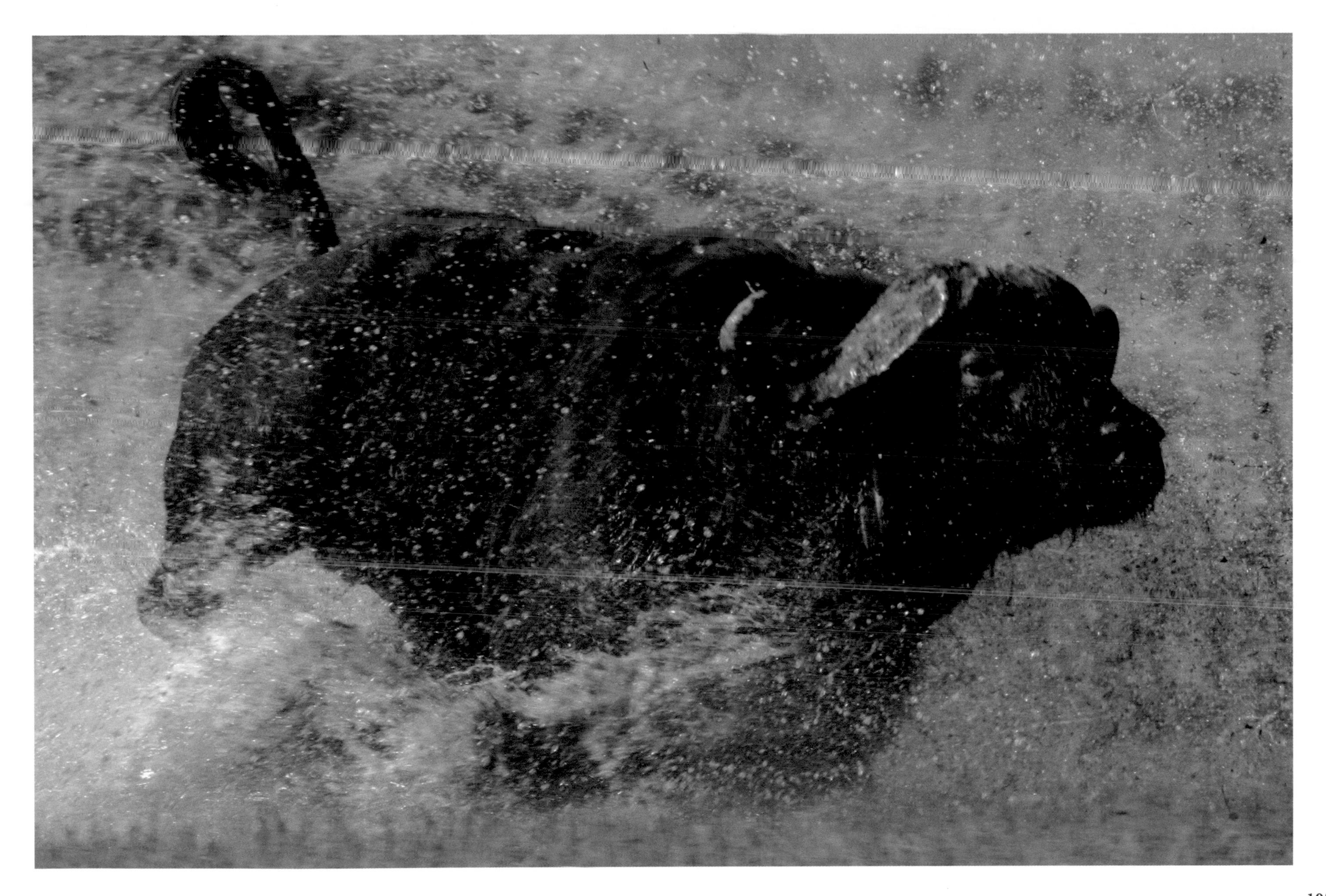

In aerial views the dance is much more evident, the tactics more easily seen.
The Tsaro females form a wall of confrontation and move in with such confidence that the herd panics.

The Kill

After witnessing a few thousand kills, we've seen enough of these events to be completely in awe of the mechanics of the kill, but not often struck by the emotions of the moment. And yet, some kills still affect us both. Some are just sad.

It is obvious that the kills that most move us are ones where we know the individual buffalo. Because of the amount of time we spend with the herd, we do get to know them quite well. Females that defend their young, and die as result, are also more difficult to watch. And it is interesting that each time we anthropomorphize the animals that we know and then see them die, we "feel" more for them. Most people who visit Africa and see our films are even more vulnerable than we are to attaching emotions to the prey and to being affected by the apparently cruel deaths these animals suffer.

While I am sure, misplaced sentiment aside, that animals do suffer and feel pain, they have their own version of it, one not exactly like ours. The buffalo will settle down to feed within minutes of a violent attack. Many will go to sleep, others ruminate on the morning's feed, but none outwardly show any signs of emotional distress. That they experience some pain and hardship is certain, though. They live a life under threat and under attack, and I imagine that, like members of our own species who perform amazing feats under stress and make stunning resistance to difficulties, they too can endure unimaginable hardships.

Suffocation is the preferred method of killing a large animal. Usually this takes the form of a throat-hold that cuts off the breathing apparatus. On a big bull, a lion may close off the nose and mouth by covering it with its own mouth. But on smaller animals, the babies and calves for example, there is little need to actually kill the animal. Lions only kill to eat, so if the animal can't get away, they eat. Only in areas with high hyena densities is it necessary to stifle the calls of the prey.

This means that quite often babies are eaten alive. It isn't for the fainthearted, but that is the way it is, and that is why we treat what we do with such reverence. When a baby buffalo keeps calling out while eight lionesses fight over which piece of its meat they want to swallow down in a rush, to fill as much of their bellies as possible before the male comes walking over to take their food, one questions oneself, the apparent immortality of our species, God, and all the possible forces that could create such suffering.

The most powerful conclusion one is left with has to be an exuberance for life while it lasts.

By the same token, though, I won't tell you that the prey goes into shock and feels nothing. There is a burden that comes with our exposure to this, to witnessing it, and I won't make it easier. While there is no doubt that a kill is violent, there is no malice involved, no intentional cruelty. Perhaps as cold a comfort as it is, that is what makes what lions do so different from what man does.

After a grueling full day's hunt and fight through the swamp, a single calf—an easy kill, perhaps—is a merciful reward for the frustrated and exhausted lions. FOLLOWING PAGES: *The hunt concludes with a masterful kill: a throat strangulation grip.*

THE INKY BLUE of the winter night gave way to a cold dawn while we drank steamy cups of rooibos tea and waited for the lions to come out of the huddle they were in. We'd found them at the change from night to dull blue. The sunrise was over fast, and the first real light fell on sleeping lions. Eight lionesses and the two remaining cubs of the Tsaro pride lay peacefully about 200 paces from the 1,200 buffalo. Just another dawn at Duba.

The buffalo dribbled off slowly from the front of the herd, and soon the herd in front of the lions had diminished to a few hundred. Ahead, the first of the buffalo were already wading through icy water. The lionesses greeted each other as they had most other days, and a few played chase games to warm up. They then walked toward the herd, and one or two of the lions dropped down into a stalk mode. This was unusual for Tsaro females, but the females behind them didn't bother much with concealment. It was a normal morning, so far.

The herd was moving right to left across the path of the lions now; most of them were in the water or across the small channel that fed a pool up ahead. I looked at the lionesses, all looking and learning from the herd, each one looking at different parts of the herd. "This is different today," I said. "Yes, they're so—intense," was Beverly's observation.

Then Beverly saw the male buffalo lying down at roughly the same time I saw the lioness to my left react, and as I quickly followed the lioness's pointer-like stance I could see that she was looking at the buffalo that Beverly had seen. Eight lions closed in through the grass on the one isolated buffalo. Each lioness had her head down, eyes focused on one thing ahead in time: the kill.

Too late the bull jumped up and skidded around to face a lioness in full flight, a missile launched from an ancient impulse. She hit his side and bounced, but another and then a third tackled his hocks and back. To our surprise he collapsed in a bundle before we could absorb anything but a flurry of action and our own impulses to trigger cameras. It was fast, but we had the scene on film.

We started up and moved a little closer, slowly, now in our hushed mode as witnesses to a death. But then we saw the two approaching bulls and a cow buffalo. They stepped forward tentatively, sniffing the air, flicking their tongues into their own nostrils.

We waited. The lions had flipped the bull over on his back, so easily even I was stunned by the swiftness and strength of the lions. I didn't notice through the lens that the "rescuers" had grown in number from three to a thousand! The whole herd was coming back, stepping forward with each groan from the upside-down bull.

The lions started to growl into their meal, tails flicking, ears back, all postures we recognized: a slow gathering of forces from the opposition, from the enemy. The lions flattened against the grass behind their kill and behind each other, as the herd advanced one collective step at a time. The front row got to within five or six paces. One bold lioness leaped an explosive defense against the buffalo, and they backed away a pace or two. Flicking tongues into those noses again, they stepped back into the space they had held a moment ago. The growls intensified and the lions' tails flicked their displeasure. Another lioness jumped forward, but the herd had learned from the last time and now stepped forward to confront her. As a lioness in the back turned and dipped her head in submission, they attacked. The buffalo ran over the bare patch of grass separating this battlefield.

The herd was enraged now, chasing lions everywhere, stomping on the fallen bull, horning him, possibly in an attempt to get him up, and spraying spit and anger across the field in answer to the growling and roaring lions. The sound closed in around us, and quite suddenly we were in their world, getting spit and blood on us as lions rushed around our vehicle chased by buffalo. Neither predators nor prey took any notice of us. It was if a window had opened, and after years of looking through glass, the obstacle, transparent as it was, had disappeared, and we were in.

Riding a storm: A Tsaro lioness battles to control a buffalo cow on the run.

Although the big bull faced his enemies with strength, their weight and violence were eventually too much for him. The seven-hour battle ended a long life that must have come close to ending many times before.

The bull is subdued by seven of the Tsaro lionesses, with a combined weight of about 2,600 pounds, nearly the weight of the buffalo.

The calls of a fallen male draw his companions back in a phalanx of slashing horns and sharp hooves that would terrify even the bravest heart.
Opposite: *The struggles for survival across these plains are an eternal battle: The buffalo fight for their right to live each moment, each breath, while the lions wrestle with their drive to survive another day's hunger.*

Diary entry May 2005

Woke up to a strange cold wind today. Wonderfully refreshing cold water crossing (not) on our way to film. Found the larger pride eyeing the lioness and her cubs across the buffalo herd. She is avoiding them all the time, but more and more they meet at the corner café . . . the buffalo herd. It feels like they are brewing for a showdown, something we hope to catch on film. We stayed with the pride, expecting them to hunt buffalo. They are still wary after getting their backsides roasted by the herd a week ago.

On this film we hired a helicopter with a stabilizing mount for perfectly stable moving images. Beverly had the same luxury. I strapped a bungee across the open door before takeoff. Only once, when she looked out at 4,500 feet, did she claw her way back inside like a cat on a high roof.

The buffalo rolled onto his belly. Another bull hooked his horns in under his and threw his head back in an attack, or something else. The effect wrenched the bull onto his feet and the herd gathered around. Within seconds he was engulfed in a mass of black jostling bodies heading for the water.

The lions rallied and ran in. The herd took flight. The bull, left behind, was again cornered and swung around against the palms, fighting and slashing his horns, desperate for life. Again, on the first bellow, the herd turned and headed back toward the stricken bull. This time their approach was less tentative. They charged in, all horns, heavy bodies, and sharp hooves, and the lions recognized the aggression and dived for safety. This time, though, the buffalo quickly escorted the bull off into the herd while the back group defended.

Then we saw the buffalo gathered around the wounded bull. He had been on the ground for at least three minutes, and in that time the lions had ripped into his soft parts. These wounds were now bleeding, and the fresh blood attracted males, females, and calves, each licking at the blood and his injuries. He braced himself under the sea of attention, and we debated whether it was the salty taste of blood that so attracted the other buffalo or something else. One idea is that it is a compassionate attentiveness; another is that the natural antiseptic in animal saliva could heal the injuries. I think it is the precious allure of salt. But the buffalo were actually quite rough and competitive in getting to lick and smell the blood. Once or twice the bull had to chase them away because he was being bumped off his feet. Then a buffalo jumped up on his back and started to mate the bull, or mock mate, because this was a female buffalo! She was shaken off only to be replaced by another buffalo, this time a male. And so it went on; while the injured bull was within the herd he was mounted and licked and bumped. When he was nudged to the outskirts, the lions were waiting.

Around midday we'd been with the bull for six hours, and apart from a few sessions of about ten minutes of rest between attacks, the lions had been alert and on the advance. They pulled the male to his knees a dozen times, only to be rebuffed by the herd. They moved across the water and into the open grassland, following, attacking, diving away, dodging buffalo charges, all the while circling around to the bull.

Finally, when we thought this stalemate would never end, a single bull stepped out of the herd. I saw him coming. As I started filming again, he walked up to the male, then put his head down and attacked the injured bull ferociously, hitting him over and over on the side and locking horns with him, finally flipping him as easily as I might lift a young child. The bull crumbled under the attack and rolled over. The attacking bull, with a shine of blood on his horns, stepped back and gave him one more head butt on the ground. Then he turned his back on the injured bull and left.

The herd followed, and within moments the Tsaro pride walked over and killed the male. It had been a seven-hour battle, and the lions were relentless. The buffalo put up a good defense that could have sent any of the lions off with their own blood covering their bodies, not just the injured bull's. In battle one would salute a noble enemy. But then they don't do this for glory. There is no Pro Patria or Purple Heart medal or ticker tape parade for battles well fought, nor do the buffalo go off to plan their counterattack one day. The lions do it to eat, and the buffalo do it to stay alive.

Up against the wall with nowhere to go, these lions have nerves of steel.

WHY THE MYSTERY BULL STEPPED OUT and put an end to it all, we will never know. My best guess may change, but today I think he had just had enough. He was irritated by the fuss and the handicap of being prevented from continuing with his day. It could have been a near-ranking bull in the herd that took advantage of the incapacitated senior to oust him from his position, or a mercy killing attempt (which may be taking it too far). It could be that he attacked simply because the bull now smelled of lions, or even more simply was behaving weirdly, and that was enough for him to want to be rid of it. Lastly, he could have been unsettled by the way this injured bull was attracting the lions. Certainly his gesture was final. The herd moved away after that and not before. And the window closed on us again. Sure, we could see through it, but as if coming out of a dream we were back in reality. The moment of intensity was over.

At least we could still see inside, but now we were analyzing and talking about the number of attacks, the motivation of the mystery bull, and suddenly a wave of sheer exhaustion washed over us both. Some of it was just the sadness of someone else's suffering. Much of what we were feeling was just the release of our concentration on getting the right focus and exposure, and in my case the variety of angles and focal lengths needed to create a moment in time that would be as unforgettable to others as it was for us. Mostly it was the exhaustion you feel when you have been intellectually challenged, made to confront your demons, and forced to think about the horror of death, your own, perhaps.

The bull had turned to face his demons time and time again, each time with a compassionless glassy eye. At every turn, everywhere he looked, the lions were there like some midnight nightmare, and each time he could have given up. He had given up, I think, at times, but some greater force within him picked him up and made him fight back. It was for nothing, of course. And what about us?

It is dark outside as I write this. A hyena stalks around camp. I go outside and look at his tracks over mine. They are small enough, but still a haunting reminder that this is sometimes, no, always, a dispassionate place, dispassionate as to whether we are here or not. Africa will survive beyond us. We may drag down some of its other inhabitants with us, like desperate swimmers in a group overwhelmed by the waves, but this place and the spirit of it will outlive us. How will we turn and face our end, I wonder? Will we judge ourselves as we do the bull, as is natural for us humans to do? No one else will. But if we do, just how will we stack up?

The kill is the pinnacle of our journey here. It is the most intense moment, a transition from the magical and athletic to a more metaphysical realization or appreciation, from life and action to death or the unknown. All transitions, especially those into the unknown and unsafe, are exciting and invigorating. So should this one be, whether it is a buffalo's, a lion's, or our own.

If we learn nothing from such an intense moment, then it is all for nothing. Only the very insensitive, those same flawed people who derive pleasure from killing a lion (for fun or "recreation") can enjoy watching the kill. All reasons other than to learn and experience some inner journey from the kill are just window dressing, avoiding the most important moment one can be exposed to, and skipping out of those lessons—a waste of time.

The day after the bull kill we didn't follow the Tsaro pride or try to find Skimmer. We went to the river and sat there for a while.

Only one lioness waited until the very last before the wall of buffalo descended to overwhelm and collect their fallen companion.

At first contact, the sheer weight of the Tsaro females is often enough to collapse a cow buffalo.

A battle of the titans, in which a large buffalo cow tries to drag her assailant off in the water, ends in defeat or discretionary retreat when the water and the threat of crocodiles overwhelm her.

In the wet season, when calves are born, the lions use a different tactic: They follow the herd and watch for single mothers left behind, the signal of the birth of a calf. The easy prey are simply picked off as they are born.

When a calf is attacked it is usually the immediate kinship group that returns to rescue it. These family groups of about 20 related buffalo are formidable and aggressive, with a powerful drive to rescue their fallen kin. Sometimes a lion's only defense is retreat.

A wounded female identified by Silver Eye is knocked down over and over between tactical retreats.
A wall of buffalo horns drives the attackers off time and again, but when the lions have tasted blood they seldom let their prey go.
From that moment the end is inevitable, despite the efforts of the fallen buffalo's kinship group.

Although we may have needed a break from the never-ending drama, the lions did not. The bull was turned into orange dust under the feet of squabbling vultures. The nine lionesses, two males, and two now-bloated cubs had eaten every scrap and were hunting again.

The dawn was filled with the scent of Duba at its best —anticipation.

A rare interaction with hyenas started the day. First they ate off part of our winch-cable housing. Toying with three lionesses, they badgered them to the point of complete frustration, forcing the lions to get up and walk away from their intended meal: the advancing buffalo. The Tsaro females, vulnerable as just a partial pride, waded out across the water away from trouble.

Suddenly the hyenas turned their attention to the buffalo, and in a bizarre turnaround, the lionesses found themselves ahead of the stampeding herd. They were ideally positioned and killed two very young calves in a dramatic spray of water that hid the calves and confused the rescue attempts of their desperate mothers. However, in what was for us a strange echo of our work at Savute, we sensed tension coming through the tree line. Shortly afterward the first drooling hyena came leaping ahead of himself in his eagerness to share in the spoils. Soon there were 13 enraged hyenas bouncing through the water, racing toward the lions. Within seconds it was over, a simple strategy of psychological aggression that overwhelmed the lions and shattered their confidence. It was one of the few hyena takeovers we'd seen at Duba. It was vicious and it was fast, almost surgical.

A kill is not always where it ends. Even after hours of pushing through water, chasing down a herd, dodging the slashing horns and thundering bodies, the lions lose their kills just about as often as they keep them. The hyenas are just one of the obstacles. The male lions are another. They could actually be considered almost parasitic on the females, if it weren't for their vigilant patrolling and protecting role.

Joseph Conrad wrote that "Action is consolatory. It is the enemy of thought and the friend of flattering illusions." Across the savannas and marshes of Duba, this "action" goes on every day, but far from being a flattering illusion, it serves to stimulate much thought in our minds.

Hyenas, demonic beasts from a lion's nightmare, are enough to strike fear in even the bravest.

Small but effective, the hyenas swoop in and take the lions' kills.
Their intimidating, aggressive approach usually sends the lions running.

It is strange that 13 hyenas are more intimidating to these lions than a thousand buffalo. Some kills just cannot be defended. The day belongs to the hyenas.

She is beautiful in her symbolic makeup, horrific as its meaning is.
Someone has died to feed her and her cubs.

Crazy day! We came across the Tsaro pride stalking something before dawn. Suddenly one lioness leapt into the grass onto something we couldn't quite make out until it stood up, a huge bull buffalo. She rode it rodeo style for a moment until the pride ran in. I'd filmed the whole scene, and although I was pleased to have got the jump on, I had no idea what lay ahead. 3 hours of battle ensued, with multiple jump ons and chaos, with crazy moments captured in high speed, and in the end, we retired from what can only be described as a battlefield, exhausted.

Our secret weapon is time. We move in and work with lions daily, until they get so used to us they hunt and kill around our vehicle as if we didn't exist.

The male and his mate, Ma di Tau, were working on their own family, a pride that was supposed to inherit his territory. The arrival of marauding lions changed all that, and the dawn of a new era was set in motion.

And in the End

This is an individual journey, for you, the buffalo, the lions, and us. We will all learn differently from it. The subject is singular and obvious. It will chase us and make us confront it, but perhaps each one's time scale is different.

On the surface, the lions have learned so much. They stalk and chase and then kill according to a new set of lion behaviors, different from what we know from other parts of Africa, even Botswana. Lions learn daily. They never turn away.

Similarly, the buffalo have learned how to survive on this island, and it is a mystery why they don't just leave. In other parts of Africa where we have worked, after a lion attack, the buffalo "bounce" to the farthest corner of their range, getting as far away as possible from the lions. Here, that would send them straight from Tsaro to Pantry to Skimmer, and they would lose three buffalo a day. So instead, after an attack the herd now just settles down nearby, often to sleep. Maybe they know it is safest where the lions have just killed, maybe it is the aftereffects of the stress they go through, but often enough the buffalo do not charge away after a kill.

They have also learned that if they are under attack the best defense is to sleep! During an ongoing attack by lions, quite often the herd just pulls together into a bundle, and slowly individuals drop down for a nap that lasts an hour or two. During that time, the lions have no option but to sleep as well; attacking a sleeping herd of buffalo all bunched together with horns facing outwards, the young hidden within, is impossible. Lions use the movement of the herd to watch for defects and weak points within the herd; these vulnerabilities are also closely guarded in the huddle of resting bodies.

In the heat, this extended rest period in the sun often tips the lions' heat exhaustion levels over the edge, while for some reason the buffalo are fine in the sun, and get up and leave the panting lions behind. These lions, however, now force themselves to get up and follow, and the equivalent of an arms race continues.

The cats are in action, regardless of the water they have to cross and the other predators they encounter.
FOLLOWING PAGES: *The lions wade through water for hours each day. This was a seven-hour day when they and we were seldom on dry land. We followed in our vehicle, adopting the philosophy that if they could do it, so could we. We were wrong.*

Soaking up each of these moments like hungry students is our way of processing what can, at times, be quite hard. The life that Beverly and I have chosen is a difficult one, often on the edge, often uncomfortable, but what we most enjoy about it is that it takes us to that meeting place of past instincts and present intellect. Only right there, at that point, are we truly alive.

None of us, though, no matter where we choose to live, can escape the primitive complexities we have inherited from our simian past, and we get involved in the intricacies of our small hardships. We sometimes draw down the blinds on the richness of each consecutive breath that automatically fills our lungs. Like a diver who has run out of air underwater once, and forevermore savors each breath, we drive through these fields littered with the bones and ghosts of the fallen and fill our souls with the reality of being alive, now. If nothing else, that should be enough.

But it isn't; our heads fill up with the pressure of being creative or just plain clever, and we study Descartes or the mathematical Spinoza, Jung and Freud, in some hope that they understood better, when in fact it is all inside and perfectly understood by that part of us that has been around for a few million years.

The relationship between the lions and the buffalo of Duba shifts and sways back and forth, changing from time to time and returning to what it was. These two ancient and relentless enemies are locked into this dance, a parallel to our own troubled relationship with nature. As a species we fight nature into apparent submission daily, but it bounces back at us. We have developed our understanding enough, however, to realize that it is the dance that is important, not who wins or loses.

And far from being disillusioned about the future of the dance, I have to confess to a great optimism. I feel the slow wind of change, just a breeze against the skin for now, but there nonetheless. Without hope, the future will be no better than the past, and we can't live with that.

The buffalo herd has done well despite the heavy predation. Numbering well over a thousand, it grows at a rate of about 5 percent per year.
Following pages: *The destiny of the Duba lions is linked to that of the buffalo; without their constant presence, here on an island with few other animals, the lions simply would not survive.*

Her enormous bulk weighs heavily on a Tsaro lioness leaping over a stream.
Opposite: *Tails paint the air with water brushstrokes as the Tsaro females fight for a piece of meat. Lions are unique in that they cooperate in the hunt but compete for the food.*

Diary entry May 2009

Today, by contrast to yesterday, when everything just happened, nothing at all happened our way. I took a hit to the knee that virtually crippled me, then a moment later the wheel twisted in a rut and nearly tore my thumb out of its socket. When we got back to camp, without a single frame, I reached into my clothing drawer and came up with a handful of shredded cloth. A dormouse has been nesting in my new shirts, well fairly new. He immediately took refuge on high ground, my head!

A dormouse is just a mouse with a fluffy tail. But that makes all the difference.
Who could be angry with a fluff ball who escapes up your leg, down one arm,
and over your cheek to a safe, high perch—and then pauses later to check out what he just ran over?

The glow of evening brings relief not only from the heat but also from the relentless hunting. Unusually, at Duba these nocturnal predators hunt at least 90 percent of their food in bright, hot daylight hours.

A new male on the island seemed familiar and very calm. Then we found his spot pattern in an old photograph, a grown male cub from the Skimmer pride. He is the future of the Tsaro legacy now. **Opposite:** *Twin sisters—near duplicates of each other—are now the core of Tsaro pride.*

Following the Lone Lioness we called Ma di Tau has been very intense. The more you focus on one individual the more involved you become, the more anxiety fills you when she is in danger, the more you weep with her loss. One such moment came late in the production.

Ma di Tau left her cubs to hunt in the water. It was a breakthrough hunt for her. The cubs could not follow in the deep water, as they would surely have drowned.

When she finally succeeded, she was exhausted. The herd of buffalo had panicked and disappeared, and she was alone. The next morning she returned to her cubs, rested, but the buffalo had been past, churning the ground into a mess of dung and sand, and making it impossible for her to track her cubs. We'd seen this before. It's a small island and buffalo often move straight past clusters of palms that hide cubs. Mothers call and call until eventually tiny faces peep out to make sure it is all clear.

Not today.

Her aching calls were deeply disturbing, and we both felt her anxiety rise after each unanswered call.

Finally there was an answer, a single answer, and we sensitively followed on as quietly as possible so as to not disturb this happy reunion.

I filmed it from a distance and then, to our shock, when she greeted the cub we saw that it had a broken back.

The scene is now famous within our small group and cutting room because we debated long and hard if it was something to even show in the film. It is heart wrenching, and in the film you can see the agony of emotion in her eyes as she looks around, then blinks long and slow, as if she understands that trying to save her cub is futile. In Africa there is a very common answer to the question of, "How are you?" It is, "Ah, we are suffering, and what can we do?" This exact sentiment seemed so apparent in Ma di Tau's eyes. She seemed to accept that deep rip in her heart or soul or wherever her emotion lies.

And yet what do we know about animal emotion? At best, I can say that we don't know anything, but it is highly unlikely that we are the only species in the universe that feels emotion. Sure, I will agree that a soldier deadened to emotion will, in the drudge of battle, turn callous and shoot a wounded enemy soldier with less emotion than he would show when putting an injured horse out of its misery. The sliding scale of emotion is difficult to fix. But the fact of emotion must exist in more than just humans.

We saw her eyes blink and watched her step away a dozen times. Each time we could hear the cub call not to be left behind again, and on each call, the lioness went back. Eventually, we watched her leave for good.

This time the cub was quiet. Some silent, sad, signal was left in the air, and yes, that is my interpretation, but I have no other. We cut the scene and sat back.

As I considered writing what we had seen there really was nothing to say, so I left it open. What is there to say over a scene that speaks universally by itself.

Filmmakers, proud pride members in many ways, we have embedded ourselves so deeply into the Tsaro pride, the life of Ma di Tau, and Duba island that sitting near lions feels like the only home we have.

A vehicle is our home, office, bedroom, kitchen, and workbench. I am very attached to our vehicle, probably because of these many hours in it, under it and dragging out of mud to safety, but sometimes we have to part company and film on foot.

Nothing was safe while Silver Eye was around, and yet later, she and Ma di Tau mended their rift and became hunting sisters again.

OPPOSITE: *Ma di Tau and her cubs. As they grew and demanded more, tension levels rose, but she never left them unless she needed to hunt.*

FOLLOWING PAGES: *And the question remains: will these last 20,000 lions survive? Will these Duba lions be some of the last?*

What You Can Do to Help

A MESSAGE FROM THE JOUBERTS: After so many years of inspiring others to work for conservation, Beverly and I looked back at our lives to see how effective we had been. The numbers speak for themselves. Inspiration alone is simply not going to do it; we have to get up and get involved. So we started the National Geographic Big Cats Initiative as an emergency fund to make a substantial difference, both in the short term and long. If you want to contribute to the fund, visit our website at www.causeanuproar.org.

- We need to raise $50 million to turn around the declining number of lions. Every dollar helps. Support the Cause an Uproar campaign by texting LIONS to 50555 to give $10.*
- You can also contribute to the Maasailand Preservation Trust's Predator Compensation Fund, through which we pay local cattle herders a fair market price when lions kill their cows, IF they don't kill the lion in return.
- In some countries, as a way to combat hunting, we purchase lion- and leopard-hunting permits and then tear them up. We can't keep shooting declining species.
- We are supporting a motion to the Convention on International Trade in Endangered Species (CITES), the international body that regulates wildlife trade, to protect lions under Appendix I, the listing of the most endangered species. Leopards (which number 50,000) and elephants (600,000) have protection, yet lions (of which there are only about 20,000) have none and need your help. Contact your representative and see if he or she can help.
- Support Senate bill 529 (S. 529), the Great Cats and Rare Canids Act of 2009, which has died in committee. Contact your representatives to find out more about it, and work with us to get it reintroduced. Its passage would pave the way for more wildlife protection.
- We need to have lions placed on the Endangered Species list in the United States, which would help prevent the importation of lion skins. Around 600 lions are shot for sport in Africa each year; of those, 556 are imported into the United States as skins. Contact National Geographic's Big Cats Initiative or Defenders of Wildlife for more information on what you can do to help.
- This is a global problem that depends on good ideas from everyone for a solution—scientists, conservationists, and people on the street. Write to us via www.causeanuproar.org.

Please join us in our many efforts to save big cats around the world.

WWW.CAUSEANUPROAR.ORG

**A $10 donation will be added to your mobile bill/deducted from prepaid balance. Msg&Data rates may apply. Text STOP to 864833 to unsubscribe. Full terms: mGive.org/T*

Will this lonely little survivor of this grand adventure be allowed to grow up, grow into a mane, and live to dominate a territory? That, as we say in the film, will depend on us.

Acknowledgments

MANY PEOPLE WORK TOGETHER TO GET A PROJECT LIKE THIS DONE.

We asked the Okavango Community Trust, the leaseholders of the Duba concession, for permission, and received it within a week. We need to thank them for their foresight in seeing the potential of this project. We would like to thank "James" 007 (Kebalibile) Piseru, guide and friend at Duba Plains safari camp, a lion expert in his own right, and everyone at Duba camp, who are our guides, managers, friends, and just good people. The directors, managers, and staff of Wilderness Safaris deserve huge thanks. When we started this film they extended hospitality with their usual grace. Then, as we fell deeper and deeper in love with Duba and I bugged them for years to let us buy in to the area and camp, they eventually folded. Thank you then extends to our partners at Great Plains, a conservation tourism company we formed together with Colin Bell, Paul Harris, Mark Read, and National Geographic. Great Plains is an attempt to enhance threatened wild places that are just too precious to see decline, and Great Plains has come to play a major part in our lives recently as well as a significant part in the conservation of Africa.

Paul De Thierry at Duba has come to our aid often enough and has been a good neighbor during these years.

The Director of Wildlife and National Parks, and, collectively, Botswana's Department of Wildlife, as well as the Minister of Environment, Kitso Mokaila, a great ambassador for wildlife in Botswana.

Of course our dear friend of many years, the present Tau e Tona (the Great Lion), President Ian Khama. Without President Khama's lifetime commitment to the wild places and wildlife in Botswana, I doubt we would have as great an abundance of them as we have today. Botswana itself would be a very different place without this great man at its helm, and when *The Economist* magazine wrote of Botswana under President Khama's leadership as the Star of Africa, they could just as convincingly have been referring to the man.

Once again we must thank the National Geographic Society and National Geographic Channels for their support, as well as Chris Johns of *National Geographic* magazine and Maura Mulvihill of Image Sales. Also, Barbara Brownell Grogan, Marianne Koszorus, Sanaa Akkach, Susan Blair, and Bridget English of the Book division.

In an ancient ritual between buffalo and lion, the battle rages relentlessly, at times in favor of one and then shifting in favor of the other in a dramatic duel of equals.

The Last Lions

Photographed by Beverly Joubert and written by Dereck Joubert

Published by the National Geographic Society

John M. Fahey, Jr., *President and Chief Executive Officer*

Gilbert M. Grosvenor, *Chairman of the Board*

Tim T. Kelly, *President, Global Media Group*

John Q. Griffin, *Executive Vice President; President, Publishing*

Nina D. Hoffman, *Executive Vice President; President, Book Publishing Group*

Prepared by the Book Division

Barbara Brownell Grogan, *Vice President and Editor in Chief*

Marianne R. Koszorus, *Director of Design*

Carl Mehler, *Director of Maps*

R. Gary Colbert, *Production Director*

Jennifer A. Thornton, *Managing Editor*

Meredith C. Wilcox, *Administrative Director, Illustrations*

Staff for This Book

Bridget A. English, *Editor*

Sanaa Akkach, *Art Director*

Susan S. Blair, *Illustrations Editor*

Judith Klein, *Production Editor*

Al Morrow, *Design Assistant*

Lindsey Smith, *Design Intern*

Manufacturing and Quality Management

Christopher A. Liedel, *Chief Financial Officer*

Phillip L. Schlosser, *Senior Vice President*

Chris Brown, *Technical Director*

Nicole Elliott, *Manager*

Rachel Faulise, *Manager*

Robert L. Barr, *Manager*

Portions of this book were published previously in the book titled *Relentless Enemies,* photographed by Beverly Joubert and written by Dereck Joubert, 2006.

The National Geographic Society is one of the world's largest nonprofit scientific and educational organizations. Founded in 1888 to "increase and diffuse geographic knowledge," the Society works to inspire people to care about the planet. National Geographic reflects the world through its magazines, television programs, films, music and radio, books, DVDs, maps, exhibitions, live events, school publishing programs, interactive media and merchandise. *National Geographic* magazine, the Society's official journal, published in English and 32 local-language editions, is read by more than 35 million people each month. The National Geographic Channel reaches 320 million households in 34 languages in 166 countries. National Geographic Digital Media receives more than 13 million visitors a month. National Geographic has funded more than 9,200 scientific research, conservation and exploration projects and supports an education program promoting geography literacy. For more information, visit nationalgeographic.com.

For more information, please call 1-800-NGS LINE (647-5463)
or write to the following address:

National Geographic Society
1145 17th Street N.W.
Washington, D.C. 20036-4688 U.S.A.

For information about special discounts for bulk purchases,
please contact National Geographic Books Special Sales: ngspecsales@ngs.org

For rights or permissions inquiries, please contact National Geographic Books Subsidiary Rights:
ngbookrights@ngs.org

ISBN: 978-1-4262-0779-2

Library of Congress Cataloging-in-Publication Data
The Library of Congress has cataloged the 2006 edition as follows:
Joubert, Beverly.
Relentless enemies : lions and buffalo / photographed by
Beverly Joubert ; and written by Dereck Joubert.
p. cm
ISBN 1-4262-0004-8
1. Lions—Botswana—Okavango Delta—Pictorial works.
2. African buffalo—Botswana—Okavango Delta—
Pictorial works. I. Joubert, Dereck. II. National Geographic
Society (U.S.) III. Title.

QL737.C23J68 2006
599.757096883—dc22

2006046160

Printed in U.S.A.

10/WOR/1